SQA

2016 SQA Pas[t Papers] With Answers

National 5

CHEMISTRY

National 5 CHEMISTRY

SQA

2014, 2015 & 2016 Exams

This book contains the official SQA 2014, 2015 and 2016 Exams for National 5 Chemistry, with associated SQA-approved answers modified from the official marking instructions that accompany the paper.

In addition the book contains study skills advice. This has been specially commissioned by Hodder Gibson, and has been written by experienced senior teachers and examiners in line with the new National 5 syllabus and assessment outlines. This is not SQA material but has been devised to provide further guidance for National 5 examinations.

Hodder Gibson is grateful to the copyright holders, as credited on the final page of the Answer Section, for permission to use their material. Every effort has been made to trace the copyright holders and to obtain their permission for the use of copyright material. Hodder Gibson will be happy to receive information allowing us to rectify any error or omission in future editions.

Hachette UK's policy is to use papers that are natural, renewable and recyclable products and made from wood grown in sustainable forests. The logging and manufacturing processes are expected to conform to the environmental regulations of the country of origin.

Orders: please contact Bookpoint Ltd, 130 Park Drive, Milton Park, Abingdon, Oxon OX14 4SE. Telephone: (44) 01235 827720. Fax: (44) 01235 400454. Lines are open 9.00–5.00, Monday to Saturday, with a 24-hour message answering service. Visit our website at www.hoddereducation.co.uk. Hodder Gibson can be contacted direct on: Tel: 0141 333 4650; Fax: 0141 404 8188; email: hoddergibson@hodder.co.uk

This collection first published in 2016 by
Hodder Gibson, an imprint of Hodder Education,
An Hachette UK Company
211 St Vincent Street
Glasgow G2 5QY

Typeset by Aptara, Inc.

Printed in the UK

A catalogue record for this title is available from the British Library

ISBN: 978-1-4718-9104-5

3 2 1

2017 2016

Introduction

Study Skills – what you need to know to pass exams!

Pause for thought

Many students might skip quickly through a page like this. After all, we all know how to revise. Do you really though?

Think about this:

"IF YOU ALWAYS DO WHAT YOU ALWAYS DO, YOU WILL ALWAYS GET WHAT YOU HAVE ALWAYS GOT."

Do you like the grades you get? Do you want to do better? If you get full marks in your assessment, then that's great! Change nothing! This section is just to help you get that little bit better than you already are.

There are two main parts to the advice on offer here. The first part highlights fairly obvious things but which are also very important. The second part makes suggestions about revision that you might not have thought about but which WILL help you.

Part 1

DOH! It's so obvious but ...

Start revising in good time

Don't leave it until the last minute – this will make you panic.

Make a revision timetable that sets out work time AND play time.

Sleep and eat!

Obvious really, and very helpful. Avoid arguments or stressful things too – even games that wind you up. You need to be fit, awake and focused!

Know your place!

Make sure you know exactly **WHEN and WHERE** your exams are.

Know your enemy!

Make sure you know what to expect in the exam.

How is the paper structured?

How much time is there for each question?

What types of question are involved?

Which topics seem to come up time and time again?

Which topics are your strongest and which are your weakest?

Are all topics compulsory or are there choices?

Learn by DOING!

There is no substitute for past papers and practice papers – they are simply essential! Tackling this collection of papers and answers is exactly the right thing to be doing as your exams approach.

Part 2

People learn in different ways. Some like low light, some bright. Some like early morning, some like evening / night. Some prefer warm, some prefer cold. But everyone uses their BRAIN and the brain works when it is active. Passive learning – sitting gazing at notes – is the most INEFFICIENT way to learn anything. Below you will find tips and ideas for making your revision more effective and maybe even more enjoyable. What follows gets your brain active, and active learning works!

Activity 1 – Stop and review

Step 1

When you have done no more than 5 minutes of revision reading STOP!

Step 2

Write a heading in your own words which sums up the topic you have been revising.

Step 3

Write a summary of what you have revised in no more than two sentences. Don't fool yourself by saying, "I know it, but I cannot put it into words". That just means you don't know it well enough. If you cannot write your summary, revise that section again, knowing that you must write a summary at the end of it. Many of you will have notebooks full of blue/black ink writing. Many of the pages will not be especially attractive or memorable so try to liven them up a bit with colour as you are reviewing and rewriting. **This is a great memory aid, and memory is the most important thing.**

Activity 2 – Use technology!

Why should everything be written down? Have you thought about "mental" maps, diagrams, cartoons and colour to help you learn? And rather than write down notes, why not record your revision material?

What about having a text message revision session with friends? Keep in touch with them to find out how and what they are revising and share ideas and questions.

Why not make a video diary where you tell the camera what you are doing, what you think you have learned and what you still have to do? No one has to see or hear it, but the process of having to organise your thoughts in a formal way to explain something is a very important learning practice.

Be sure to make use of electronic files. You could begin to summarise your class notes. Your typing might be slow, but it will get faster and the typed notes will be easier to read than the scribbles in your class notes. Try to add different fonts and colours to make your work stand out. You can easily Google relevant pictures, cartoons and diagrams which you can copy and paste to make your work more attractive and **MEMORABLE**.

Activity 3 – This is it. Do this and you will know lots!

Step 1

In this task you must be very honest with yourself! Find the SQA syllabus for your subject (www.sqa.org.uk). Look at how it is broken down into main topics called MANDATORY knowledge. That means stuff you MUST know.

Step 2

BEFORE you do ANY revision on this topic, write a list of everything that you already know about the subject. It might be quite a long list but you only need to write it once. It shows you all the information that is already in your long-term memory so you know what parts you do not need to revise!

Step 3

Pick a chapter or section from your book or revision notes. Choose a fairly large section or a whole chapter to get the most out of this activity.

With a buddy, use Skype, Facetime, Twitter or any other communication you have, to play the game "If this is the answer, what is the question?". For example, if you are revising Geography and the answer you provide is "meander", your buddy would have to make up a question like "What is the word that describes a feature of a river where it flows slowly and bends often from side to side?".

Make up 10 "answers" based on the content of the chapter or section you are using. Give this to your buddy to solve while you solve theirs.

Step 4

Construct a wordsearch of at least 10 × 10 squares. You can make it as big as you like but keep it realistic. Work together with a group of friends. Many apps allow you to make wordsearch puzzles online. The words and phrases can go in any direction and phrases can be split. Your puzzle must only contain facts linked to the topic you are revising. Your task is to find 10 bits of information to hide in your puzzle, but you must not repeat information that you used in Step 3. DO NOT show where the words are. Fill up empty squares with random letters. Remember to keep a note of where your answers are hidden but do not show your friends. When you have a complete puzzle, exchange it with a friend to solve each other's puzzle.

Step 5

Now make up 10 questions (not "answers" this time) based on the same chapter used in the previous two tasks. Again, you must find NEW information that you have not yet used. Now it's getting hard to find that new information! Again, give your questions to a friend to answer.

Step 6

As you have been doing the puzzles, your brain has been actively searching for new information. Now write a NEW LIST that contains only the new information you have discovered when doing the puzzles. Your new list is the one to look at repeatedly for short bursts over the next few days. Try to remember more and more of it without looking at it. After a few days, you should be able to add words from your second list to your first list as you increase the information in your long-term memory.

FINALLY! Be inspired...

Make a list of different revision ideas and beside each one write **THINGS I HAVE** tried, **THINGS I WILL** try and **THINGS I MIGHT** try. Don't be scared of trying something new.

And remember – "FAIL TO PREPARE AND PREPARE TO FAIL!"

National 5 Chemistry

The course

Before sitting your National 5 Chemistry examination, you must have passed three **Unit Assessments** within your school or college, and produced an additional short report (approximately 100 words).

To achieve a pass in National 5 Chemistry there are then two further main components.

Component 1 – Assignment

You are required to submit an assignment that is worth 20% (20 marks) of your final grade. This assignment will be based on research and may include an experiment. This assignment requires you to apply skills, knowledge and understanding to investigate a relevant topic in chemistry and its effect on the environment and/or society. Your school or college will provide you with a Candidate's Guide for this assignment, which has been produced by the SQA. This guide gives guidance on what is required to complete the 400–800 word report and gain as many marks as possible.

Your assignment report will be marked by the SQA.

Component 2 – The Question Paper

The question paper will assess breadth and depth of knowledge and understanding from across all of the three Units. The question paper will require you to:

- Make statements, provide explanations, and describe information to demonstrate knowledge and understanding.
- Apply knowledge and understanding to new situations to solve problems.
- Plan and design experiments.
- Present information in various forms such as graphs, tables etc.
- Perform calculations based on information given.
- Give predictions or make generalisations based on information given.
- Draw conclusions based on information given.
- Suggest improvement to experiments to improve the accuracy of results obtained or to improve safety.

To achieve a "C" grade in National 5 Chemistry you must achieve about 50% of the 100 marks available when the two components, i.e. the Question Paper and the Assignment are combined. For a B you will need 60%, while for an "A" grade you must ensure that you gain as many of the available marks as possible, and at least 70%.

Each SQA Past Paper consists of two sections. (A marking scheme for each section is provided at the end of this book.)

- Section 1 will contain objective questions (multiple choice) and will have 20 marks.
- Section 2 will contain restricted and extended response questions and will have 60 marks.

Each SQA Past Paper contains a variety of questions including some that require:

- demonstration and application of knowledge, and understanding of the mandatory content of the course from across the three units
- application of scientific inquiry skills.

How to use this book

This book can be used in two ways:

1. You can complete an entire paper under exam conditions, without the use of books or notes, and then mark the papers using the marking scheme provided. This method gives you a clear indication of the level you are working at and should highlight the content areas that you need to work on before attempting the next paper. This method also allows you to see your progress as you complete each paper.
2. You can complete a paper using your notes and books. Try the question first and then refer to your notes if you are unable to answer the question. This is a form of studying and by doing this you will cover all the areas of content that you are weakest in. You should notice that you are referring to your notes less with each paper completed.

Try to practise as many questions as possible. This will get you used to the language used in the papers and ultimately improve your chances of success.

Some hints and tips

Below is a list of hints and tips that will help you to achieve your full potential in the National 5 exam.

- Ensure that you **read each question carefully**. Scanning the question and missing the main points results in mistakes being made. Some students highlight the main points of a question with a highlighter pen to ensure that they don't miss anything out.
- Open ended questions include the statement **"Using your knowledge of chemistry"**. These questions provide you with an opportunity to show off your chemistry knowledge. To obtain the

three marks on offer for these questions, you must demonstrate a good understanding of the chemistry involved and provide a logically correct answer to the question posed.

- When doing calculations, ensure that you **show all of your working**. If you make a simple arithmetical mistake you may still be awarded some of the marks, but only if your working is laid out clearly so that the examiner can see where you went wrong and what you did correctly. Just giving the answers is very risky so you should always show your working.
- **Attempt all questions.** Giving no answer at all means that you will definitely not gain any marks.
- When you are required to read a passage to answer a question, ensure that you **read it carefully** as the information you require is contained within it. It may not be obvious at first, but the answers will be contained within the passage.
- If you are asked to "explain" in a question, then you must **explain your answer fully**. For example, if you are asked to explain how a covalent bond holds atoms together then you cannot simply say:

  "A covalent bond is a shared pair of electrons between atoms in a non-metal."

  This answer tells the examiner what a covalent bond is, but does not explain how it holds the atoms together. To gain the marks, an answer similar to this should be written:

  "A covalent bond is a shared pair of electrons between atoms in a non-metal. The shared electrons are attracted to the nuclei of both atoms, which creates a tug-of-war effect creating the covalent bond."
- You will be required to draw one graph in each exam. To obtain all the marks, ensure that the graphs have **labels, units, points plotted correctly** and a line of "best fit" drawn between the points.
- Use your **data booklet** when you are asked to write formulas, ionic formulas, formula mass etc. You have the data booklet in front of you so use it to double check the numbers you require.
- Work on your **timing**. The multiple-choice section (Section 1) should take approximately 30 minutes. Attempt to answer the multiple-choice questions before you look at the four possible answers, as this will improve your confidence. Use scrap paper when required to scribble down structural formulae, calculations, chemical formulae etc., as this will reduce your chance of making errors. If you are finding the question difficult, try to eliminate the obviously wrong answers to increase your chances.
- When asked to **predict or estimate** based on information from a graph or a table, then take your time to look for patterns. For example, if asked to predict a boiling point, try to establish if there is a regular change in boiling point and use that regular pattern to establish the unknown boiling point.
- When drawing a **diagram** of an experiment ask yourself the question, "Would this work if I set it up exactly like this in the lab?" Ensure that the method you have drawn would produce the desired results safely. If, for example, you are heating a flammable reactant such as alcohol then you will not gain the marks if you heat it with a Bunsen burner in your diagram; a water bath would be much safer! Make sure your diagram is labelled clearly.

Good luck!

Remember that the rewards for passing National 5 Chemistry are well worth it! Your pass will help you get the future you want for yourself. In the exam, be confident in your own ability. If you're not sure how to answer a question, trust your instincts and just give it a go anyway. Keep calm and don't panic! GOOD LUCK!

NATIONAL 5

2014

X713/75/02

Chemistry
Section 1—Questions

MONDAY, 12 MAY

9:00 AM—11:00 AM

Necessary data will be found in the Chemistry Data Booklet for National 5.

Instructions for the completion of Section 1 are given on *Page two* of your question and answer booklet X713/75/01.

Record your answers on the answer grid on *Page three* of your question and answer booklet

Before leaving the examination room you must give your question and answer booklet to the Invigilator; if you do not, you may lose all the marks for this paper.

SECTION 1

1. In a reaction, 60 cm^3 of hydrogen gas was collected in 20 s.

   What is the average rate of reaction, in $cm^3\,s^{-1}$, over this time?

   A $\frac{60}{20}$

   B $\frac{20}{60}$

   C $\frac{1}{60}$

   D $\frac{1}{20}$

2. Molecules in which four different atoms are attached to a carbon atom are said to be chiral.

   Which of the following molecules is chiral?

   A
```
      Br
      |
      C
    /   \
   H  |  H
      Cl
```

   B
```
      Br
      |
      C
    /   \
   H  |  H
      H
```

   C
```
      I
      |
      C
    /   \
  Cl  |  H
      H
```

   D
```
      I
      |
      C
    /   \
   H  |  Br
      F
```

3. What is the charge on the zinc ion in the compound zinc phosphate $Zn_3(PO_4)_2$?

A 2+

B 3+

C 2−

D 3−

4. $Fe_2O_3 + \mathbf{x}\,CO \longrightarrow \mathbf{y}\,Fe + 3CO_2$

This equation will be balanced when

A **x** = 1 and **y** = 2

B **x** = 2 and **y** = 2

C **x** = 3 and **y** = 2

D **x** = 2 and **y** = 3.

5. An acidic solution contains

A only hydrogen ions

B only hydroxide ions

C more hydrogen ions than hydroxide ions

D more hydroxide ions than hydrogen ions.

6. Which of the following oxides, when shaken with water, would give an alkaline solution?

A Calcium oxide

B Nickel oxide

C Nitrogen dioxide

D Sulfur dioxide

7. Which of the following compounds is **not** a salt?

A Calcium nitrate

B Sodium chloride

C Potassium sulfate

D Magnesium hydroxide

8. $H^+(aq) + NO_3^-(aq) + K^+(aq) + OH^-(aq) \longrightarrow K^+(aq) + NO_3^-(aq) + H_2O(\ell)$

The spectator ions present in the reaction above are

A $K^+(aq)$ and $NO_3^-(aq)$

B $K^+(aq)$ and $H^+(aq)$

C $OH^-(aq)$ and $NO_3^-(aq)$

D $H^+(aq)$ and $OH^-(aq)$.

9. The molecular formula for cyclohexane is

A C_6H_6

B C_6H_{10}

C C_6H_{12}

D C_6H_{14}.

10.

```
      H   H   H   H
      |   |   |   |
  H — C — C — C — C — H
      |   |   |   |
      H   H   |   H
              |
          H — C — H
              |
              H
```

The systematic name for the structure shown is

A 1,1-dimethylpropane

B 2-methylbutane

C 3-methylbutane

D 2-methylpentane.

11. Petrol is a mixture of hydrocarbons.

The tendency of a hydrocarbon to ignite spontaneously is measured by its octane number.

	Hydrocarbon	*Octane number*
1	3-methylpentane	74·5
2	butane	93·6
3	pentane	61·7
4	2-methylpentane	73·4
5	hexane	24·8
6	methylcyclopentane	91·3

A student made the hypothesis that as the chain length of a hydrocarbon increases, the octane number decreases.

Which set of three hydrocarbons should have their octane numbers compared in order to test this hypothesis?

A 1, 4, 6

B 1, 2, 4

C 2, 3, 5

D 3, 4, 5

12. Propene reacts with hydrogen bromide to form two products.

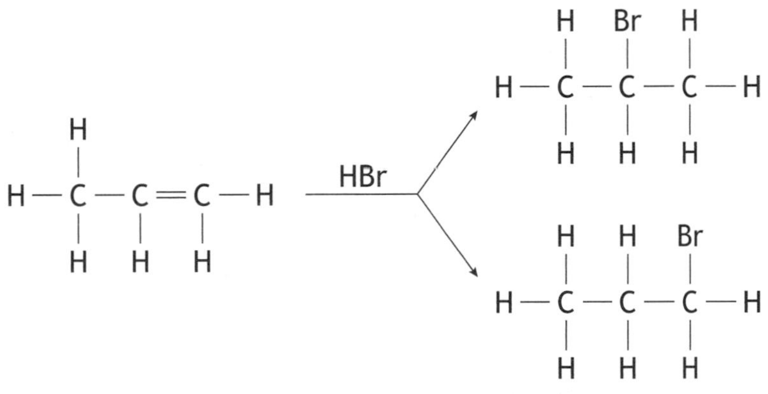

```
    H                                  H   Br  H
    |                                  |   |   |
H — C — C = C — H      HBr       H  —  C — C — C — H
    |   |   |                          |   |   |
    H   H   H                          H   H   H

                                       H   H   Br
                                       |   |   |
                                 H  —  C — C — C — H
                                       |   |   |
                                       H   H   H
```

Which of the following alkenes does **not** form two products on reaction with hydrogen bromide?

A But-1-ene

B But-2-ene

C Pent-1-ene

D Pent-2-ene

13. Which of the following alcohols has the highest boiling point?

You may wish to use your data booklet to help you.

A Propan-1-ol

B Propan-2-ol

C Butan-1-ol

D Butan-2-ol

14. A reaction is endothermic if

A energy is required to start the reaction

B heat is released during the reaction

C the temperature drops during the reaction

D the temperature rises during the reaction.

15. Which of the following metals will **not** react with a dilute solution of hydrochloric acid?

A Copper

B Iron

C Magnesium

D Zinc

16. Which metal can be extracted from its oxide by heat alone?

A Tin

B Zinc

C Lead

D Silver

17. The ion-electron equations for the oxidation and reduction steps in the reaction between **sulfite ions** and **iron(III) ions** are given below.

oxidation $H_2O(\ell) + SO_3^{2-}(aq) \longrightarrow SO_4^{2-}(aq) + 2H^+(aq) + 2e^-$

reduction $Fe^{3+}(aq) + e^- \longrightarrow Fe^{2+}(aq)$

The redox equation for the overall reaction is

A $H_2O(\ell) + SO_3^{2-}(aq) + Fe^{3+}(aq) \longrightarrow SO_4^{2-}(aq) + 2H^+(aq) + Fe^{2+}(aq) + e^-$

B $H_2O(\ell) + SO_3^{2-}(aq) + 2Fe^{3+}(aq) \longrightarrow SO_4^{2-}(aq) + 2H^+(aq) + 2Fe^{2+}(aq)$

C $SO_4^{2-}(aq) + 2H^+(aq) + Fe^{2+}(aq) + e^- \longrightarrow H_2O(\ell) + SO_3^{2-}(aq) + Fe^{3+}(aq)$

D $SO_4^{2-}(aq) + 2H^+(aq) + 2Fe^{2+}(aq) \longrightarrow H_2O(\ell) + SO_3^{2-}(aq) + 2Fe^{3+}(aq)$.

18. The apparatus below was set up.

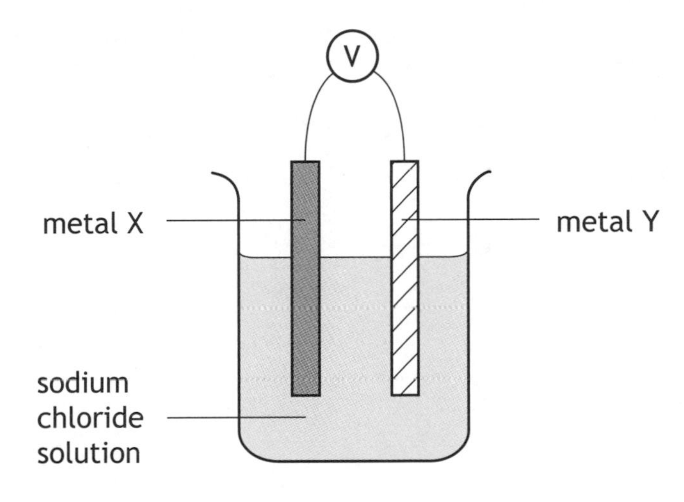

Which of the following pairs of metals would give the highest reading on the voltmeter?

	Metal X	Metal Y
A	Iron	Zinc
B	Magnesium	Silver
C	Zinc	Copper
D	Zinc	Silver

[Turn over

19. A section of a condensation polymer is shown below.

$$-\overset{\overset{\displaystyle O}{\|}}{C}-C_6H_4-\overset{\overset{\displaystyle O}{\|}}{C}-O-(CH_2)_2-O-\overset{\overset{\displaystyle O}{\|}}{C}-C_6H_4-\overset{\overset{\displaystyle O}{\|}}{C}-$$

One of the monomers is

$$H-O-\overset{\overset{\displaystyle O}{\|}}{C}-C_6H_4-\overset{\overset{\displaystyle O}{\|}}{C}-O-H$$

The structural formula for the other monomer is

A $H-\overset{\overset{\displaystyle O}{\|}}{C}-O-(CH_2)_2-O-\overset{\overset{\displaystyle O}{\|}}{C}-H$

B $H-O-(CH_2)_2-O-H$

C $H-O-\overset{\overset{\displaystyle O}{\|}}{C}-(CH_2)_2-O-H$

D $H-O-\overset{\overset{\displaystyle O}{\|}}{C}-(CH_2)_2-\overset{\overset{\displaystyle O}{\|}}{C}-O-H$

20. $Ba^{2+}(aq) + 2NO_3^-(aq) + 2Na^+(aq) + SO_4^{2-}(aq) \longrightarrow Ba^{2+}SO_4^{2-}(s) + 2Na^+(aq) + 2NO_3^-(aq)$

The type of reaction represented by the equation above is

A addition

B displacement

C neutralisation

D precipitation.

[END OF SECTION 1. NOW ATTEMPT THE QUESTIONS IN SECTION 2 OF YOUR QUESTION AND ANSWER BOOKLET]

X713/75/01

Chemistry
Section 1—Answer Grid And Section 2

MONDAY, 12 MAY

9:00 AM—11:00 AM

Fill in these boxes and read what is printed below.

Full name of centre

Town

Forename(s)

Surname

Number of seat

Date of birth

Day | Month | Year

D D | M M | Y Y

Scottish candidate number

Necessary Data will be found in the Chemistry Data Booklet for National 5.

Total marks — 80

SECTION 1 — 20 marks

Attempt ALL questions in this section.

Instructions for the completion of Section 1 are given on *Page two*.

SECTION 2 — 60 marks

Attempt ALL questions in this section.

Write your answers clearly in the spaces provided in this booklet. Additional space for answers and rough work is provided at the end of this booklet. If you use this space you must clearly identify the question number you are attempting. Any rough work must be written in this booklet. You should score through your rough work when you have written your final copy.

Use **blue** or **black** ink.

Before leaving the examination room you must give this booklet to the Invigilator; if you do not, you may lose all the marks for this paper.

SECTION 1 — 20 marks

The questions for Section 1 are contained in the question paper X713/75/02.
Read these and record your answers on the answer grid on *Page three* opposite.
Do NOT use gel pens.

1. The answer to each question is **either** A, B, C or D. Decide what your answer is, then fill in the appropriate bubble (see sample question below).

2. There is **only one correct** answer to each question.

3. Any rough work must be written in the additional space for answers and rough work at the end of this booklet.

Sample Question

To show that the ink in a ball-pen consists of a mixture of dyes, the method of separation would be

A fractional distillation

B chromatography

C fractional crystallisation

D filtration.

The correct answer is **B**—chromatography. The answer **B** bubble has been clearly filled in (see below).

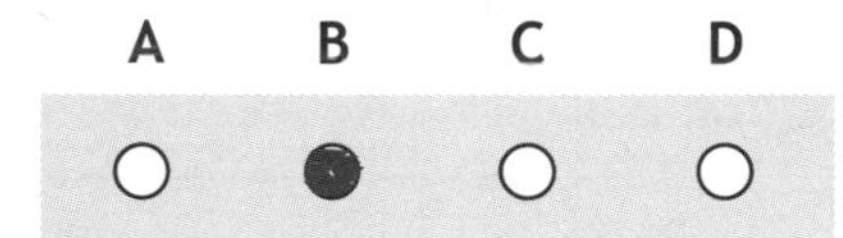

Changing an answer

If you decide to change your answer, cancel your first answer by putting a cross through it (see below) and fill in the answer you want. The answer below has been changed to **D**.

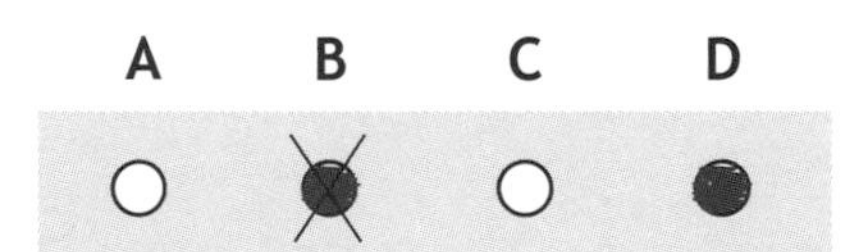

If you then decide to change back to an answer you have already scored out, put a tick (✓) to the **right** of the answer you want, as shown below:

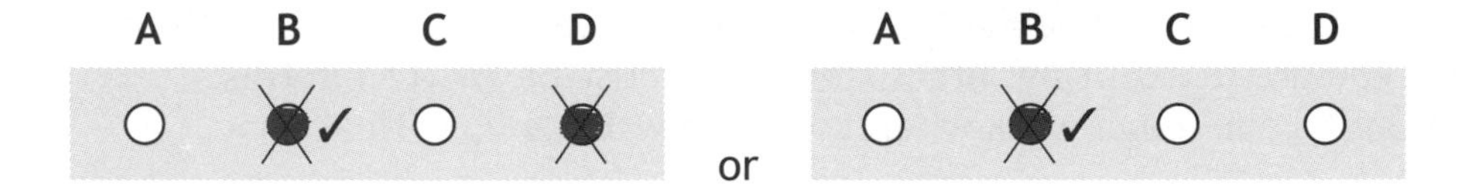

SECTION 1 — Answer Grid

	A	B	C	D
1	○	○	○	○
2	○	○	○	○
3	○	○	○	○
4	○	○	○	○
5	○	○	○	○
6	○	○	○	○
7	○	○	○	○
8	○	○	○	○
9	○	○	○	○
10	○	○	○	○
11	○	○	○	○
12	○	○	○	○
13	○	○	○	○
14	○	○	○	○
15	○	○	○	○
16	○	○	○	○
17	○	○	○	○
18	○	○	○	○
19	○	○	○	○
20	○	○	○	○

[Turn over

[BLANK PAGE]

DO NOT WRITE ON THIS PAGE

[Turn over for Question 1 on *Page six*

DO NOT WRITE ON THIS PAGE

SECTION 2 — 60 marks
Attempt ALL questions

MARKS | DO NOT WRITE IN THIS MARGIN

1. In 1911, Ernest Rutherford carried out an experiment to confirm the structure of the atom. In this experiment, he fired positive particles at a very thin layer of gold foil. Most of the particles passed straight through but a small number of the positively charged particles were deflected.

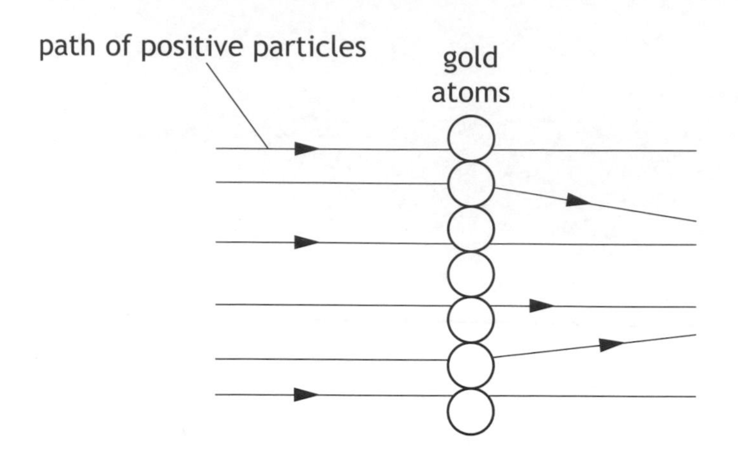

(a) What caused some of the positive particles to be deflected in this experiment? **1**

(b) Gold is the heaviest element to have only one naturally occurring isotope.

The isotope has a mass number of 197.

(i) Complete the table to show the number of each type of particle in this gold atom. **1**

You may wish to use the data booklet to help you.

Particle	*Number*
Proton	
Electron	
Neutron	

(ii) Most elements have more than one isotope.

State what is meant by the term isotope. **1**

Total marks 3

MARKS | DO NOT WRITE IN THIS MARGIN

2. (a) The properties of a substance depend on its type of bonding and structure. There are four types of bonding and structure.

Discrete covalent molecular	Covalent network	Ionic lattice	Metallic lattice

Complete the table to match up each type of bonding and structure with its properties.

Type of bonding and structure	*Properties*
	do not conduct electricity and have high melting points
	have high melting points and conduct electricity when liquid but not when solid
	conduct electricity when solid and have a wide range of melting points
	do not conduct electricity and have low melting points

2

(b) Graphene is a substance made of a single layer of carbon atoms.

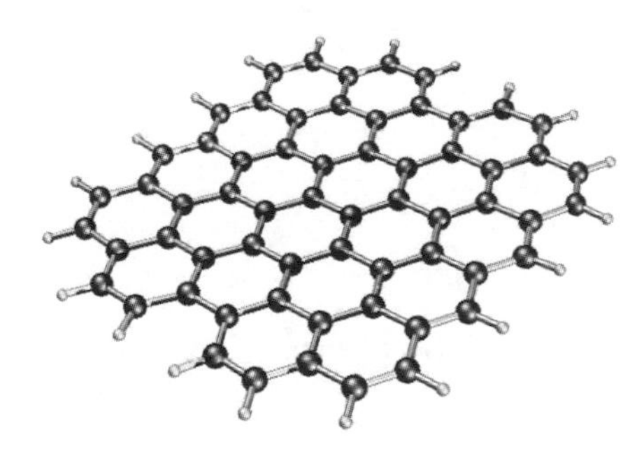

Graphene can conduct electricity.

Suggest what this indicates about some of the electrons in graphene. 1

Total marks 3

[Turn over

MARKS | DO NOT WRITE IN THIS MARGIN

3. Read the passage below and answer the questions that follow.

Potassium — The Super Element

Potassium is an essential element for almost all living things. The human body requires a regular intake of potassium because humans have no mechanism for storing it. Foods rich in potassium include raisins and almonds. Raisins contain 0·86 g of potassium in every 100 g.

Naturally occurring salts of potassium such as saltpetre (potassium nitrate) and potash (potassium carbonate) have been known for centuries. Potassium salts are used as fertilisers.

Potassium was first isolated by Humphry Davy in 1807. Davy observed that when potassium was added to water it formed globules which skimmed about on the surface, burning with a coloured flame and forming an alkaline solution.

(a) State why the human body requires a regular intake of potassium. **1**

(b) Calculate the number of moles of potassium in 100 g of raisins. **2**

Show your working clearly.

(c) State the colour of the flame which would be seen when potassium burns. **1**

You may wish to use the data booklet to help you.

(d) Write the **ionic** formula for saltpetre. **1**

Total marks 5

MARKS | DO NOT WRITE IN THIS MARGIN

4. Poly(vinylcarbazole) is a plastic which conducts electricity when exposed to light.

The structure of the monomer used to make poly(vinylcarbazole) is

$$\begin{array}{ccc} NC_{12}H_8 & & H \\ | & & | \\ C & = & C \\ | & & | \\ H & & H \end{array}$$

(a) Draw a section of the polymer showing three monomer units joined together. **1**

(b) Name the type of polymerisation taking place when these monomers join together. **1**

Total marks 2

[Turn over

MARKS | DO NOT WRITE IN THIS MARGIN

5. Different types of radiation have different penetrating properties.

An investigation was carried out using three radioactive sources.

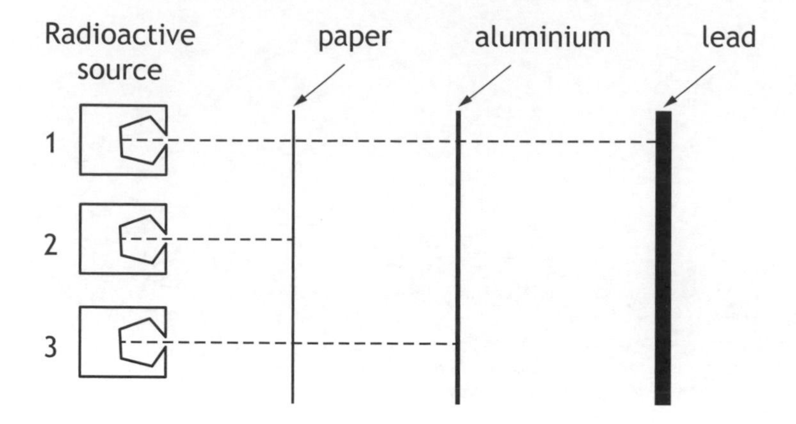

(a) Name the type of radiation emitted by source **2**. **1**

(b) The half-life of source **3** is 8 days.

Calculate the fraction of source **3** that would remain after 16 days. **2**

Show your working clearly.

(c) Radioisotopes can be made by scientists.

The nuclear equation shows how a radioisotope of element **X** can be made from aluminium.

$$^{27}_{13}Al + ^{1}_{0}n \longrightarrow X + ^{4}_{2}He$$

Name element **X**. **1**

Total marks 4

MARKS DO NOT WRITE IN THIS MARGIN

6. A student reacted acidified potassium permanganate solution with oxalic acid, $C_2H_2O_4$.

$$2MnO_4^-(aq) + 5C_2H_2O_4(aq) + 6H^+(aq) \longrightarrow 2Mn^{2+}(aq) + 10CO_2(g) + 8H_2O(\ell)$$

Using your knowledge of chemistry, comment on how the student could have determined the rate of the reaction. 3

[Turn over

MARKS DO NOT WRITE IN THIS MARGIN

7. The manufacture of potassium nitrate, for use in fertilisers, can be split into three stages.

(a) (i) In stage **1**, ammonia is produced.

Name the industrial process used to manufacture ammonia. **1**

(ii) Draw a diagram to show how **all** the outer electrons are arranged in a molecule of ammonia, NH_3. **1**

(b) In stage **2**, ammonia is converted into nitric acid, HNO_3, as shown in the flow diagram.

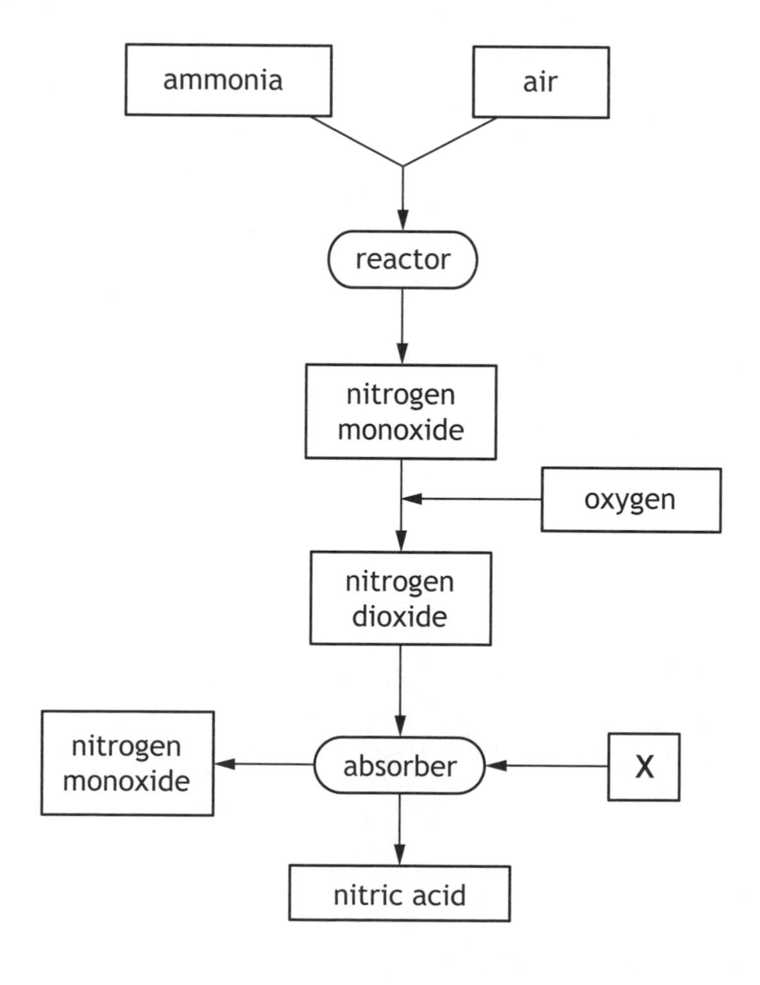

MARKS | DO NOT WRITE IN THIS MARGIN

7. (b) (continued)

(i) Name substance **X**. 1

(ii) **On the flow diagram**, draw an arrow to show how the process can be made more economical. 1

(c) In stage **3**, nitric acid is converted to potassium nitrate.

The equation for the reaction taking place is

$$HNO_3(aq) \quad + \quad KOH(aq) \quad \longrightarrow \quad KNO_3(aq) \quad + \quad H_2O(\ell)$$

(i) Name the type of chemical reaction taking place in stage **3**. 1

(ii) State how a sample of **solid** potassium nitrate could be obtained from the potassium nitrate solution. 1

Total marks 6

[Turn over

MARKS | DO NOT WRITE IN THIS MARGIN

8. Pheromones are chemicals, produced by living things, that trigger a response in members of the same species.

When a bee stings an animal the bee also releases a pheromone containing the ester below.

```
                       H
                       |
                    H— C — H
        O      H   H   |   H
        ||     |   |   |   |
  H     C      C — C — C — C — H
   \   / \   / |   |   |   |
     C     O   H   H   H   H
   /   \
  H     H
```

(a) State another use for esters. **1**

(b) A student made the ester above using ethanoic acid and the following alcohol.

```
          H
          |
     H— C — H
     H    |    H   H
     |    |    |   |
 H — C — C — C — C — O — H
     |    |    |   |
     H    H    H   H
```

(i) Name the functional group present in this alcohol. **1**

(ii) Draw a structural formula for an isomer of this alcohol. **1**

MARKS | DO NOT WRITE IN THIS MARGIN

8. (b) (continued)

(iii) Ethanoic acid is the second member of a family of compounds which contain the carboxyl functional group.

The full structural formulae for the first three members of this family are shown.

```
        O              H     O              H   H     O
       //              |    //              |   |    //
  H — C          H — C — C            H — C — C — C
       \               |    \               |   |    \
        O — H          H     O — H          H   H     O — H
```

methanoic acid | ethanoic acid | propanoic acid

Suggest a general formula for this family of compounds. **1**

(c) The table gives information on some other esters.

Alcohol	*Carboxylic acid*	*Ester*
methanol	ethanoic acid	methyl ethanoate
propanol	methanoic acid	propyl methanoate
butanol	ethanoic acid	butyl ethanoate
pentanol	butanoic acid	pentyl butanoate
X	Y	ethyl propanoate

Name **X** and **Y**. **2**

Total marks 6

[Turn over

MARKS DO NOT WRITE IN THIS MARGIN

9. Liquefied petroleum gas (LPG), which can be used as a fuel for heating, is a mixture of propane and butane.

(a) Propane and butane are members of the homologous series of alkanes.

Tick (✓) the **two** boxes that correctly describe members of the same homologous series. **1**

	Tick (✓)
They have similar chemical properties.	
They have the same molecular formula.	
They have the same general formula.	
They have the same physical properties.	
They have the same formula mass.	

(b) The table gives some information about propane and butane.

Alkane	*Boiling Point* (°C)
propane	−42
butane	−1

Explain why butane has a higher boiling point than propane. **2**

MARKS DO NOT WRITE IN THIS MARGIN

9. (continued)

(c) 25 kg of water at 10 °C is heated by burning some LPG.

Calculate the energy, in kJ, required to increase the temperature of the water to 30 °C. **3**

You may wish to use the data booklet to help you.

Show your working clearly.

(d) LPG is odourless. In order to detect gas leaks, ethyl mercaptan, C_2H_6S, a smelly gas, is added in small quantities to the LPG mixture.

Suggest one disadvantage of adding sulfur compounds, such as ethyl mercaptan, to fuels such as LPG. **1**

Total marks 7

MARKS DO NOT WRITE IN THIS MARGIN

10. The lowest temperature at which a hydrocarbon ignites is called its flash point.

Hydrocarbon	*Flash point* (°C)
hexane	−23
heptane	−4
octane	13
nonane	31

(a) (i) Using the information in the table, make a general statement linking the flash point to the number of carbon atoms. **1**

(ii) Predict the flash point, in °C, of decane, $C_{10}H_{22}$. **1**

MARKS DO NOT WRITE IN THIS MARGIN

10. (continued)

(b) Nonane burns to produce carbon dioxide and water.

$$C_9H_{20} \quad + \quad 14O_2 \longrightarrow 9CO_2 \quad + \quad 10H_2O$$

Calculate the mass, in grams, of carbon dioxide produced when 32 g of nonane is burned. **3**

Show your working clearly.

Total marks 5

[Turn over

MARKS | DO NOT WRITE IN THIS MARGIN

11. Chlorine can be produced commercially from concentrated sodium chloride solution in a membrane cell. Only sodium ions can pass through the membrane. These ions move in the direction shown in the diagram.

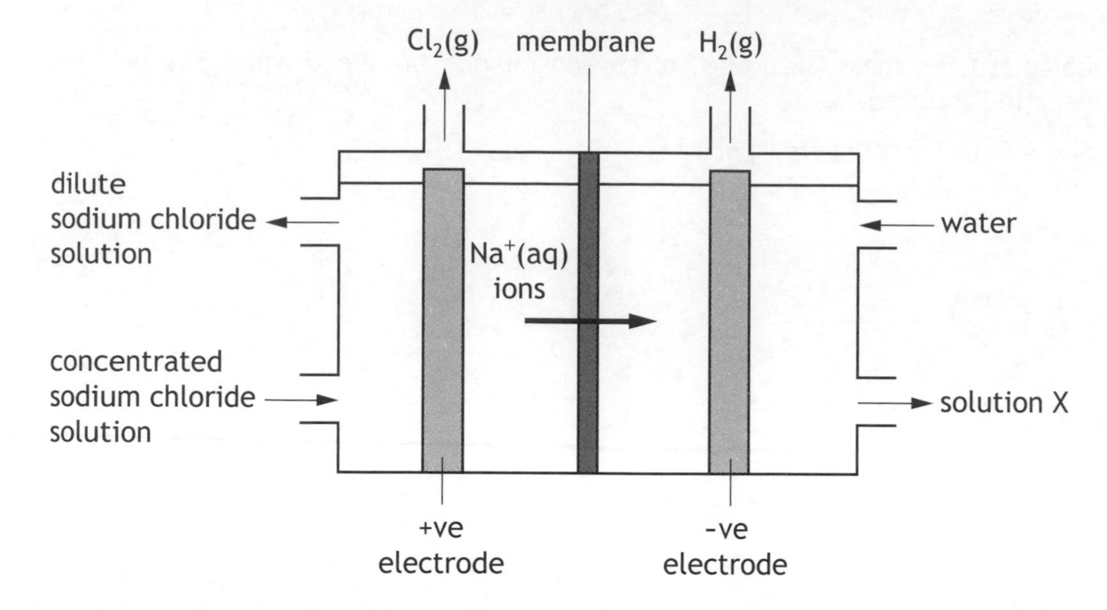

(a) Write the ion-electron equation for the change taking place at the positive electrode. **1**

You may wish to use the data booklet to help you.

(b) (i) Name solution **X**. **1**

(ii) The hydrogen gas produced, at the negative electrode, can be used as a fuel.

Suggest an advantage of using hydrogen as a fuel. **1**

MARKS | DO NOT WRITE IN THIS MARGIN

11. (continued)

(c) The chlorine gas produced can be used to make phosgene, $COCl_2$. Phosgene is used in the manufacture of drugs and plastics.

Draw a possible structure for phosgene. **1**

Total marks 4

[Turn over

MARKS | DO NOT WRITE IN THIS MARGIN

12. Ores are naturally occurring compounds from which metals can be extracted.

(a) When a metal is extracted from its ore, metal ions are changed to metal atoms.

Name this type of chemical reaction. **1**

(b) Iron can be extracted from its ore haematite, Fe_2O_3, in a blast furnace.

Calculate the percentage by mass of iron in haematite. **3**

Show your working clearly.

(c) Magnesium cannot be extracted from its ore in a blast furnace.

Suggest a method that would be suitable for the extraction of magnesium from its ore. **1**

Total marks 5

MARKS | DO NOT WRITE IN THIS MARGIN

13. Sodium carbonate solution can be added to the water in swimming pools to neutralise the acidic effects of chlorine.

A student carried out a titration experiment to determine the concentration of a sodium carbonate solution.

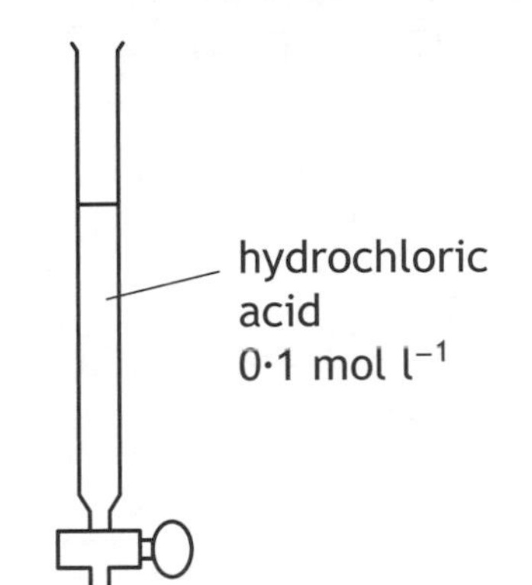

	Rough titre	*1st titre*	*2nd titre*
Initial burette reading (cm^3)	0·0	0·0	0·0
Final burette reading (cm^3)	16·5	15·9	16·1
Volume used (cm^3)	16·5	15·9	16·1

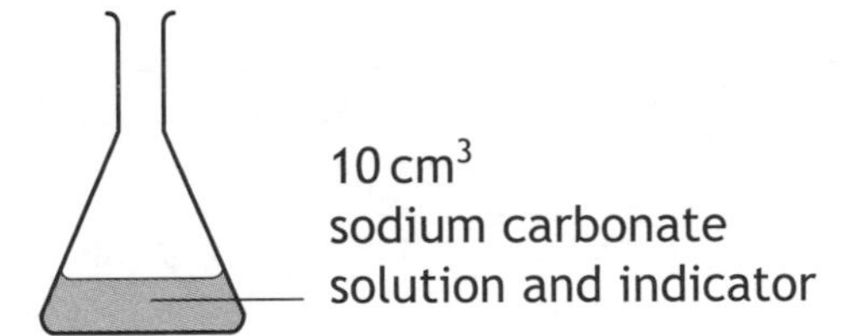

(a) Using the results in the table, calculate the average volume, in cm^3, of hydrochloric acid required to neutralise the sodium carbonate solution. **1**

(b) The equation for the reaction is

$$2HCl + Na_2CO_3 \longrightarrow 2NaCl + CO_2 + H_2O$$

Using your answer from part (a) calculate the concentration, in mol l^{-1}, of the sodium carbonate solution. **3**

Show your working clearly.

Total marks 4

MARKS | DO NOT WRITE IN THIS MARGIN

14. Chemistry in the cinema.

In the film Dante's Peak, a family trapped by red hot lava escape by crossing a large lake in a boat made from aluminium. The volcano releases heat and the gases hydrogen chloride, sulfur dioxide and sulfur trioxide into the water in the lake. While crossing the lake, holes begin to appear in the bottom of the boat. Just after the family leave the boat, on the other side of the lake, the boat sinks.

Using your knowledge of chemistry, comment on whether or not the events described in the film could take place. **3**

[END OF QUESTION PAPER]

MARKS DO NOT WRITE IN THIS MARGIN

ADDITIONAL SPACE FOR ANSWERS AND ROUGH WORK

MARKS | DO NOT WRITE IN THIS MARGIN

ADDITIONAL SPACE FOR ANSWERS AND ROUGH WORK

NATIONAL 5

2015

X713/75/02

Chemistry
Section 1—Questions

THURSDAY, 28 MAY

9:00 AM—11:00 AM

Instructions for the completion of Section 1 are given on *Page two* of your question and answer booklet X713/75/01.

Record your answers on the answer grid on *Page three* of your question and answer booklet.

Necessary data will be found in the Chemistry Data Booklet for National 5.

Before leaving the examination room you must give your question and answer booklet to the Invigilator; if you do not, you may lose all the marks for this paper.

SECTION 1

1. An atom has 26 protons, 26 electrons and 30 neutrons.

The atom has

A atomic number 26, mass number 56

B atomic number 56, mass number 30

C atomic number 30, mass number 26

D atomic number 52, mass number 56.

2. The table shows the numbers of protons, electrons and neutrons in four particles, **W**, **X**, **Y** and **Z**.

Particle	*Protons*	*Electrons*	*Neutrons*
W	17	17	18
X	11	11	12
Y	17	17	20
Z	18	18	18

Which pair of particles are isotopes?

A **W** and **X**

B **W** and **Y**

C **X** and **Y**

D **Y** and **Z**

3. Which of the following particles contains a different number of electrons from the others?

You may wish to use the data booklet to help you.

A Cl^-

B S^{2-}

C Ar

D Na^+

4. Which of the following diagrams shows the apparatus which would allow a soluble gas to be removed from a mixture of gases?

A

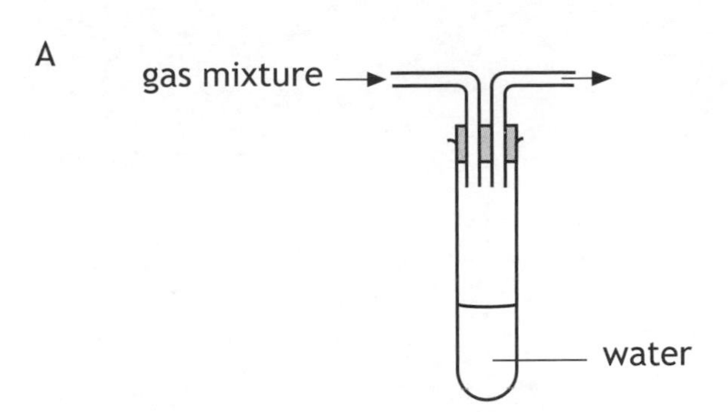

B

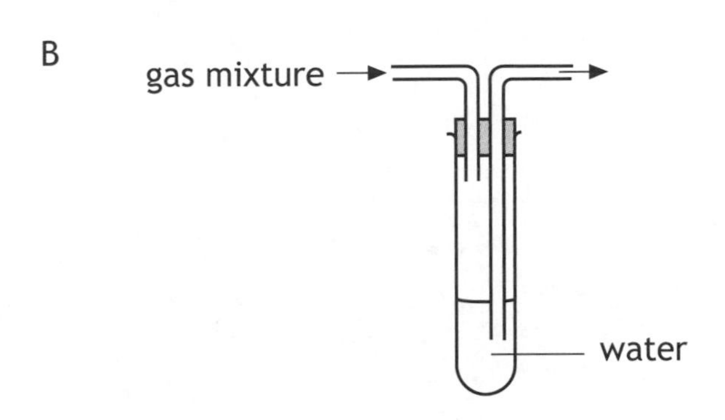

C

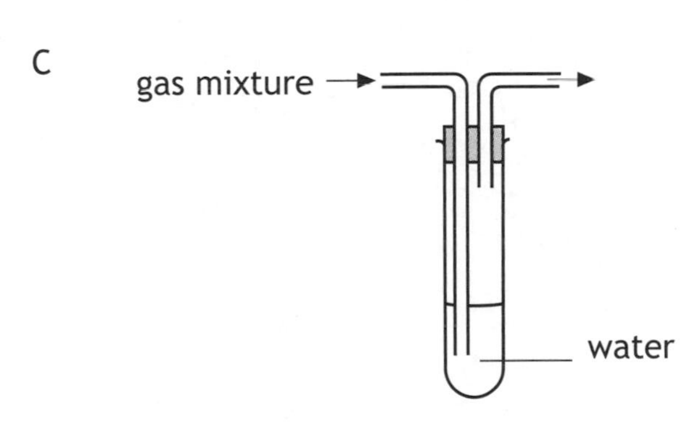

D

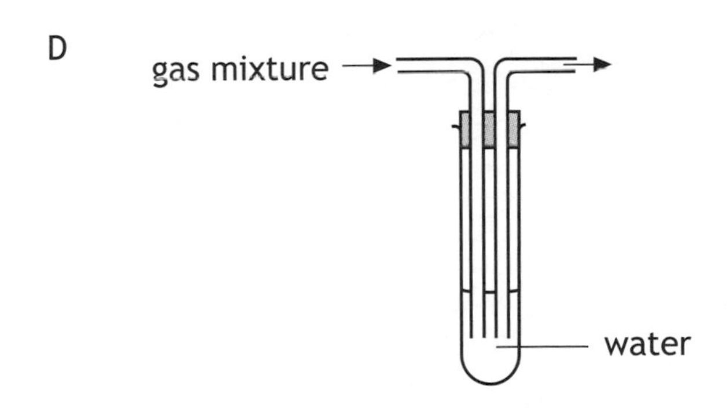

[Turn over

5. Which of the following diagrams could be used to represent the structure of a covalent network?

A

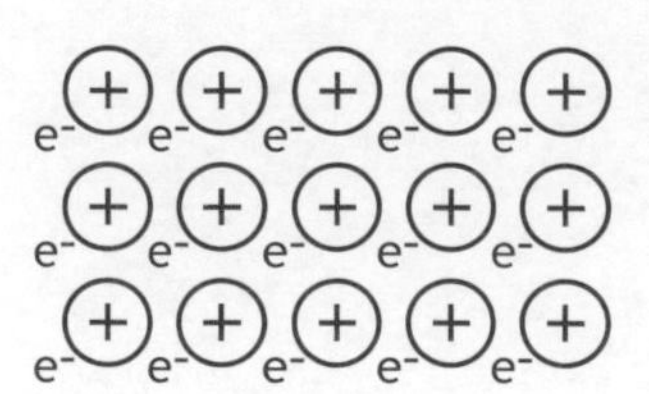

B

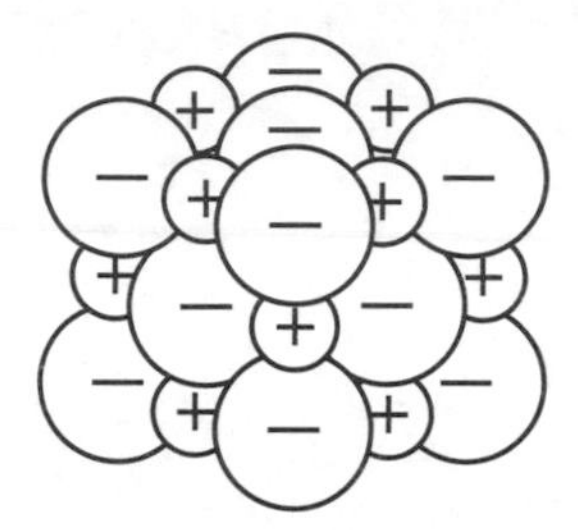

C

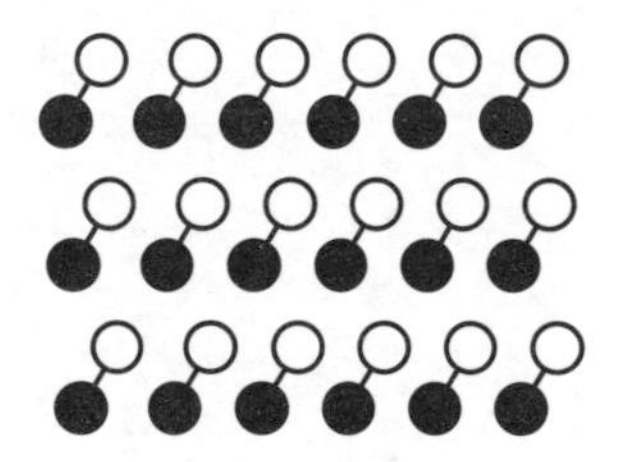

D

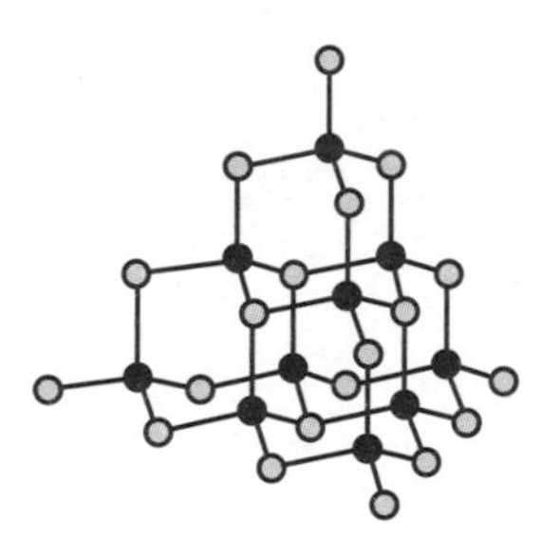

6. What is the charge on the chromium ion in $CrCl_3$?

A 1+

B 1−

C 3+

D 3−

7. The table contains information about calcium and calcium chloride.

	Melting point (°C)	*Density* ($g\,cm^{-3}$)
Calcium	842	1·54
Calcium chloride	772	2·15

When molten calcium chloride is electrolysed at 800 °C the calcium appears as a

A solid at the bottom of the molten calcium chloride

B liquid at the bottom of the molten calcium chloride

C solid on the surface of the molten calcium chloride

D liquid on the surface of the molten calcium chloride.

8. $\mathbf{x}\,Al(s) + \mathbf{y}\,Br_2(\ell) \rightarrow \mathbf{z}\,AlBr_3(s)$

This equation will be balanced when

A $\mathbf{x} = 1$, $\mathbf{y} = 2$ and $\mathbf{z} = 1$

B $\mathbf{x} = 2$, $\mathbf{y} = 3$ and $\mathbf{z} = 2$

C $\mathbf{x} = 3$, $\mathbf{y} = 2$ and $\mathbf{z} = 3$

D $\mathbf{x} = 4$, $\mathbf{y} = 3$ and $\mathbf{z} = 4$.

9. 0·2 mol of a gas has a mass of 12·8 g.

Which of the following could be the molecular formula for the gas?

A SO_2

B CO

C CO_2

D NH_3

[Turn over

10. Which of the following oxides, when shaken with water, would leave the pH unchanged?

You may wish to use the data booklet to help you.

A Carbon dioxide

B Copper oxide

C Sodium oxide

D Sulfur dioxide

11. Which compound would **not** neutralise hydrochloric acid?

A Sodium carbonate

B Sodium chloride

C Sodium hydroxide

D Sodium oxide

12.

```
     H   H   H   CH3 H
     |   |   |   |   |
H —  C — C — C — C — C — H
     |   |   |   |   |
     H   H  CH3  H   H
```

The name of the above compound is

A 2,3-dimethylpropane

B 3,4-dimethylpropane

C 2,3-dimethylpentane

D 3,4-dimethylpentane.

13. The shortened structural formula for an organic compound is

$CH_3CH(CH_3)CH(OH)C(CH_3)_3$

Which of the following is another way of representing this structure?

A

```
     H   H   OH  CH3
     |   |   |   |
 H — C — C — C — C — CH3
     |   |   |   |
     H  CH3  H   CH3
```

B

```
     H   H   H   OH  CH3
     |   |   |   |   |
 H — C — C — C — C — C — CH3
     |   |   |   |   |
     H   H   H   H   CH3
```

C

```
     H   H   H   CH3 CH3
     |   |   |   |   |
 H — C — C — C — C — C — H
     |   |   |   |   |
     H  CH3  OH  H   H
```

D

```
     H   H   H   H   H   H
     |   |   |   |   |   |
 H — C — C — C — C — C — C — CH3
     |   |   |   |   |   |
     H  CH3  OH  H   H   H
```

[Turn over

14. Three members of the cycloalkene homologous series are

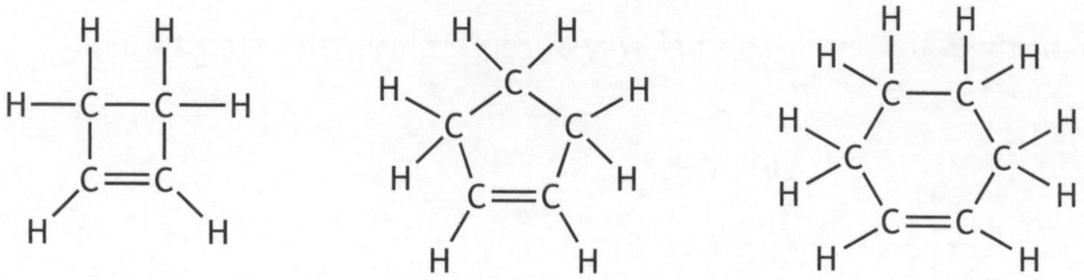

Which of the following is the general formula for this homologous series?

A C_nH_{2n-4}

B C_nH_{2n+2}

C C_nH_{2n}

D C_nH_{2n-2}

15. Metallic bonding is a force of attraction between

A negative ions and positive ions

B a shared pair of electrons and two nuclei

C positive ions and delocalised electrons

D negative ions and delocalised electrons.

16. Which pair of metals, when connected in a cell, would give the highest voltage and a flow of electrons from **X** to **Y**?

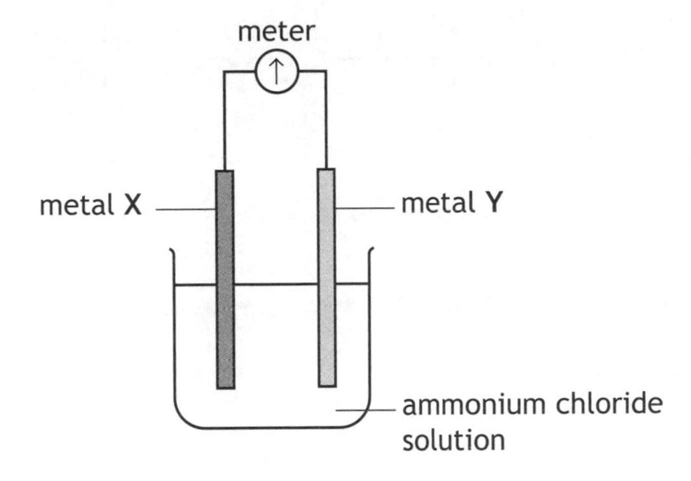

You may wish to use the data booklet to help you.

	Metal X	*Metal Y*
A	zinc	tin
B	tin	zinc
C	copper	magnesium
D	magnesium	copper

[Turn over

17. Part of the structure of a polymer is drawn below.

```
   H    H    H    H    H    H
   |    |    |    |    |    |
 —C —— C —— C —— C —— C —— C—
   |    |    |    |    |    |
  CH3   H   CH3   H   CH3   H
```

The monomer used to make this polymer is

A
```
  H    H
  |    |
  C == C
  |    |
 CH3   H
```

B
```
    H    H
    |    |
 —— C —— C ——
    |    |
   CH3   H
```

C
```
  H    H    H
  |    |    |
  C == C —— C —— H
  |         |
 CH3       CH3
```

D
```
    H    H    H
    |    |    |
 —— C —— C —— C ——
    |    |    |
   CH3   H   CH3
```

18. Sodium sulfate solution reacts with barium chloride solution.

$$Na_2SO_4(aq) \quad + \quad BaCl_2(aq) \longrightarrow BaSO_4(s) \quad + \quad 2NaCl(aq)$$

The spectator ions present in this reaction are

A Na^+ and Cl^-

B Na^+ and SO_4^{2-}

C Ba^{2+} and Cl^-

D Ba^{2+} and SO_4^{2-}.

19. Which of the following solutions would produce a precipitate when mixed together?

You may wish to use the data booklet to help you.

A Ammonium chloride and potassium nitrate

B Zinc nitrate and magnesium sulfate

C Calcium nitrate and nickel chloride

D Sodium iodide and silver nitrate

[Turn over for Question 20 on *Page twelve*

20. The table shows the colours of some ionic compounds in solution.

Compound	*Colour*
copper sulfate	blue
copper chromate	green
potassium chloride	colourless
potassium chromate	yellow

The colour of the chromate ion is

A blue

B green

C colourless

D yellow.

[END OF SECTION 1. NOW ATTEMPT THE QUESTIONS IN SECTION 2 OF YOUR QUESTION AND ANSWER BOOKLET]

FOR OFFICIAL USE

National
Qualifications
2015

Mark

X713/75/01

Chemistry
Section 1—Answer Grid And Section 2

THURSDAY, 28 MAY

9:00 AM—11:00 AM

Fill in these boxes and read what is printed below.

Full name of centre

Town

Forename(s)

Surname

Number of seat

Date of birth

Day Month Year

Scottish candidate number

Total marks — 80

SECTION 1 — 20 marks

Attempt ALL questions.

Instructions for the completion of Section 1 are given on *Page two*.

SECTION 2 — 60 marks

Attempt ALL questions.

Necessary Data will be found in the Chemistry Data Booklet for National 5.

Write your answers clearly in the spaces provided in this booklet. Additional space for answers and rough work is provided at the end of this booklet. If you use this space you must clearly identify the question number you are attempting. Any rough work must be written in this booklet. You should score through your rough work when you have written your final copy.

Use **blue** or **black** ink.

Before leaving the examination room you must give this booklet to the Invigilator; if you do not, you may lose all the marks for this paper.

SECTION 1 — 20 marks

The questions for Section 1 are contained in the question paper X713/75/02.
Read these and record your answers on the answer grid on *Page three* opposite.
Use **blue** or **black** ink. Do NOT use gel pens or pencil.

1. The answer to each question is **either** A, B, C or D. Decide what your answer is, then fill in the appropriate bubble (see sample question below).

2. There is **only one correct** answer to each question.

3. Any rough work must be written in the additional space for answers and rough work at the end of this booklet.

Sample Question

To show that the ink in a ball-pen consists of a mixture of dyes, the method of separation would be

A fractional distillation

B chromatography

C fractional crystallisation

D filtration.

The correct answer is **B**—chromatography. The answer **B** bubble has been clearly filled in (see below).

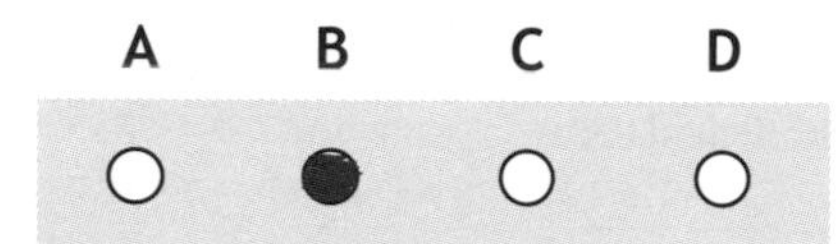

Changing an answer

If you decide to change your answer, cancel your first answer by putting a cross through it (see below) and fill in the answer you want. The answer below has been changed to **D**.

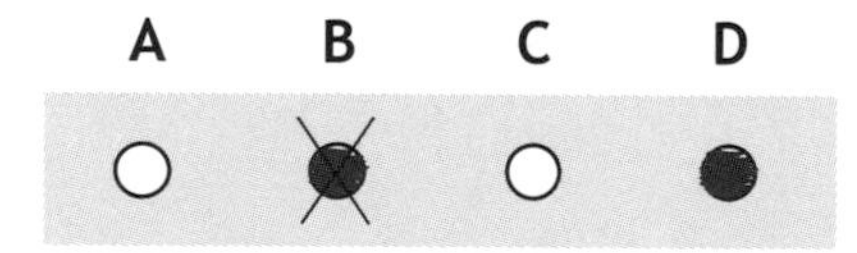

If you then decide to change back to an answer you have already scored out, put a tick (✓) to the **right** of the answer you want, as shown below:

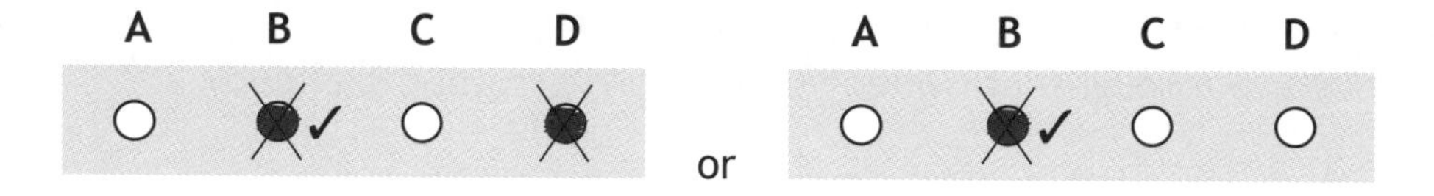

SECTION 1 — Answer Grid

	A	B	C	D
1	○	○	○	○
2	○	○	○	○
3	○	○	○	○
4	○	○	○	○
5	○	○	○	○
6	○	○	○	○
7	○	○	○	○
8	○	○	○	○
9	○	○	○	○
10	○	○	○	○
11	○	○	○	○
12	○	○	○	○
13	○	○	○	○
14	○	○	○	○
15	○	○	○	○
16	○	○	○	○
17	○	○	○	○
18	○	○	○	○
19	○	○	○	○
20	○	○	○	○

[Turn over

[BLANK PAGE]

DO NOT WRITE ON THIS PAGE

[Turn over for Question 1 on *Page six*

DO NOT WRITE ON THIS PAGE

MARKS | DO NOT WRITE IN THIS MARGIN

SECTION 2 — 60 marks

Attempt ALL questions

1. Ethyne is the first member of the alkyne family.

   It can be produced by the reaction of calcium carbide with water.

   The equation for this reaction is

$$CaC_2(s) \quad + \quad 2H_2O(\ell) \longrightarrow C_2H_2(g) \quad + \quad Ca(OH)_2(aq)$$

   (a) The table shows the results obtained in an experiment carried out to measure the volume of ethyne gas produced.

Time (s)	0	30	60	90	120	150	180	210
Volume of ethyne (cm^3)	0	60	96	120	140	148	152	152

   Calculate the average rate of reaction between 60 and 90 seconds.

   Your answer must include the appropriate unit. **3**

   Show your working clearly.

MARKS | DO NOT WRITE IN THIS MARGIN

1. (continued)

(b) Draw a line graph of the results.

Use appropriate scales to fill most of the graph paper. **3**

(Additional graph paper, if required, will be found on *Page twenty-seven.*)

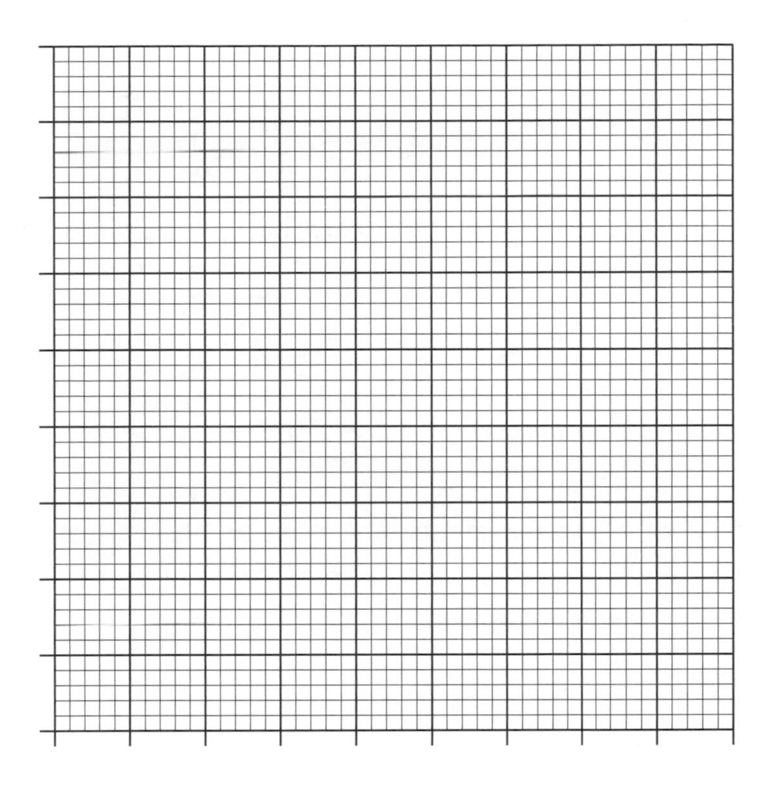

[Turn over

MARKS DO NOT WRITE IN THIS MARGIN

2. Americium-241, a radioisotope used in smoke detectors, has a half-life of 432 years.

(a) The equation for the decay of americium-241 is

$$^{241}_{95}Am \longrightarrow {}^{4}_{2}He + X$$

Name element **X**. 1

(b) Name the **type** of radiation emitted by the americium-241 radioisotope. 1

(c) Another radioisotope of americium exists which has an atomic mass of 242.

Americium-242 has a half-life of 16 hours.

(i) A sample of americium-242 has a mass of 8 g.

Calculate the mass, in grams, of americium-242 that would be left after 48 hours. 2

Show your working clearly.

(ii) Suggest why americium-241, and not americium-242, is the radioisotope used in smoke detectors. 1

MARKS | DO NOT WRITE IN THIS MARGIN

3. Butter contains different triglyceride molecules.

(a) A triglyceride molecule is made when the alcohol glycerol reacts with carboxylic acids.

(i) Name the functional group present in glycerol. **1**

(ii) Name the family to which triglycerides belong. **1**

(b) When butter goes off, a triglyceride molecule is broken down, producing compounds **X** and **Y**.

$$HO-\overset{\overset{\displaystyle O}{\|}}{C}-C_3H_7 \qquad HO-\overset{\overset{\displaystyle O}{\|}}{C}-C_{17}H_{33}$$

X Y

(i) Name compound **X**. **1**

(ii) Describe the chemical test, including the result, to show that compound **Y** is unsaturated. **1**

[Turn over

MARKS DO NOT WRITE IN THIS MARGIN

4. Some sources of methane gas contain hydrogen sulfide, H_2S.

(a) Draw a diagram, showing all outer electrons, to represent a molecule of hydrogen sulfide, H_2S. **1**

(b) If hydrogen sulfide is not removed before methane gas is burned, sulfur dioxide is formed.

When sulfur dioxide dissolves in water in the atmosphere, acid rain is produced.

Circle the correct words to complete the sentence.

Acid rain contains more {hydrogen / hydroxide} ions than {hydrogen / hydroxide} ions. **1**

(c) In industry, calcium oxide is reacted with sulfur dioxide to reduce the volume of sulfur dioxide released into the atmosphere.

Explain why calcium oxide is able to reduce the volume of sulfur dioxide gas released. **2**

MARKS | DO NOT WRITE IN THIS MARGIN

5. A researcher investigated the conditions for producing ammonia.

$$N_2(g) + 3H_2(g) \rightleftharpoons 2NH_3(g)$$

(a) Name the catalyst used in the production of ammonia. **1**

(b) In her first experiment she measured how the percentage yield of ammonia varied with pressure at a constant temperature of 500 °C.

Pressure (atmospheres)	100	200	300	400	500
Percentage yield (%)	10	18	26	32	40

Predict the percentage yield of ammonia at 700 atmospheres. **1**

(c) In a second experiment the researcher kept the pressure constant, at 200 atmospheres, and changed the temperature as shown.

Temperature (°C)	200	300	400	500
Percentage yield (%)	89	67	39	18

Describe how the percentage yield varies with temperature. **1**

(d) **Using the information in both tables**, suggest the combination of temperature and pressure that would produce the highest percentage yield of ammonia. **1**

[Turn over

MARKS | DO NOT WRITE IN THIS MARGIN

6. Read the passage below and answer the questions that follow.

Clean coal technology comes a step closer

It is claimed a process called Coal-Direct Chemical Looping (CDCL) is able to release energy from coal while capturing 99% of the carbon dioxide emitted. CDCL works by extracting the energy from coal using a reaction other than combustion.

A mixture of powdered coal and beads of iron(III) oxide is heated inside a metal cylinder. Carbon in the coal and oxygen from the beads react to form carbon dioxide which can be captured for recycling or stored.

This reaction gives off heat energy that could be used to heat water in order to drive electricity-producing steam turbines.

Adapted from *Focus: Science and Technology*, April 2013

(a) The CDCL process produced 300 tonnes of carbon dioxide.

Calculate the mass, in tonnes, of carbon dioxide released into the atmosphere. **1**

(b) Write the ionic formula for the iron compound used in CDCL. **1**

(c) State the term used to describe all chemical reactions that release heat energy. **1**

MARKS DO NOT WRITE IN THIS MARGIN

7. A student was asked to carry out an experiment to determine the concentration of a copper(II) sulfate solution.

Part of the work card used is shown.

Determination of the Concentration of Copper(II) Sulfate Solution

1. Weigh an empty crucible
2. Add 100 cm^3 copper(II) sulfate solution
3. Evaporate the solution to dryness
4. Weigh the crucible containing dry copper(II) sulfate

(a) Suggest how the student could have evaporated the solution to dryness. **1**

(b) The student found that the 100 cm^3 solution contained 3·19 g of copper(II) sulfate, $CuSO_4$.

Calculate the concentration of the solution in mol l^{-1}. **2**

Show your working clearly.

[Turn over

MARKS | DO NOT WRITE IN THIS MARGIN

8. A student calculated the energy absorbed by water when ethanol is burned using two different methods.

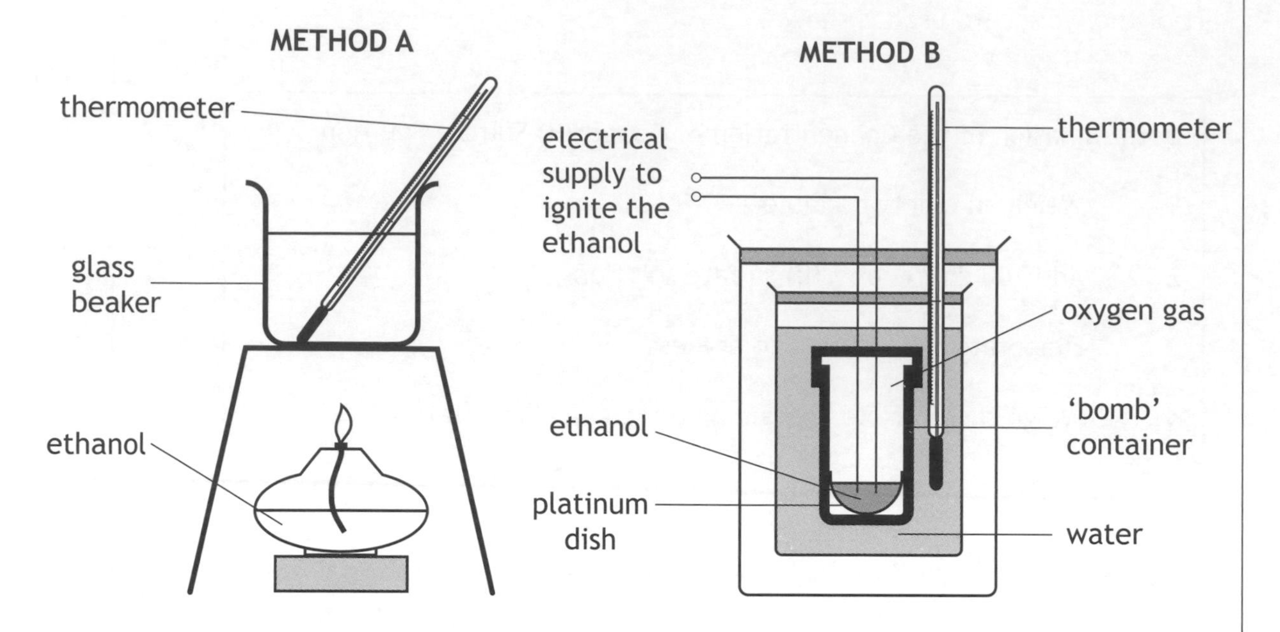

The student recorded the following data.

	Method	
	A	B
Mass of ethanol burned (g)	0·5	0·5
Mass of water heated (g)	100	100
Initial temperature of water (°C)	24	24
Final temperature of water (°C)	32	58

(a) The final temperature of water in method **B** is higher than in method **A**.

Suggest why there is a difference in the energy absorbed by the water. **1**

MARKS DO NOT WRITE IN THIS MARGIN

8. (continued)

(b) Calculate the energy, in kJ, absorbed by the water in method **B**. 3

You may wish to use the data booklet to help you.

Show your working clearly.

[Turn over

MARKS DO NOT WRITE IN THIS MARGIN

9. Aluminium can be extracted from naturally occurring metal compounds such as bauxite.

(a) State the term used to describe naturally occurring metal compounds such as bauxite. **1**

(b) Bauxite is refined to produce aluminium oxide.

Electrolysis of molten aluminium oxide produces aluminium and oxygen gas.

The ion-electron equations taking place during the electrolysis of aluminium oxide are

$$Al^{3+} + 3e^- \longrightarrow Al$$

$$2O^{2-} \longrightarrow O_2 + 4e^-$$

(i) Write the redox equation for the overall reaction. **1**

(ii) State why ionic compounds, like aluminium oxide, conduct electricity when molten. **1**

MARKS DO NOT WRITE IN THIS MARGIN

9. (continued)

(c) Bauxite contains impurities such as silicon dioxide.

Silicon can be extracted from silicon dioxide as shown.

$$SiO_2 \quad + \quad 2Mg \longrightarrow Si \quad + \quad 2MgO$$

Identify the reducing agent in this reaction. **1**

[Turn over

MARKS DO NOT WRITE IN THIS MARGIN

10. A group of students were given strips of aluminium, iron, tin and zinc.

Using your knowledge of chemistry, suggest how the students could identify each of the four metals. 3

MARKS | DO NOT WRITE IN THIS MARGIN

11. Electrons can be removed from all atoms.

The energy required to do this is called the ionisation energy.

The first ionisation energy for an element is defined as the energy required to remove one mole of electrons from one mole of atoms, in the gaseous state.

The equation for the first ionisation energy of chlorine is

$$Cl(g) \longrightarrow Cl^{+}(g) + e^{-}$$

(a) State the electron arrangement for the Cl^{+} ion. **1**

You may wish to use the data booklet to help you.

(b) Write the equation for the first ionisation energy of magnesium. **1**

(c) Information on the first ionisation energy of some elements is given in the table.

Element	*First ionisation energy* ($kJ\ mol^{-1}$)
lithium	526
fluorine	1690
sodium	502
chlorine	1260
potassium	425
bromine	1150

Describe the trend in the first ionisation energy going down a group in the Periodic Table. **1**

MARKS | DO NOT WRITE IN THIS MARGIN

12. The structural formulae of two hydrocarbons are shown.

```
                                   H   H
    H           H                  |   |
    |           |              H — C — C — H
H — C — C = C — C — H              |   |
    |   |   |   |              H — C — C — H
    H   H   H   H                  |   |
                                   H   H

          A                          B
```

(a) Name hydrocarbon **A**. 1

(b) Hydrocarbons **A** and **B** can be described as isomers.

State what is meant by the term isomer. 1

MARKS DO NOT WRITE IN THIS MARGIN

12. (continued)

(c) Hydrocarbon **A** can undergo an addition reaction with water to form butan-2-ol as shown.

```
    H           H
    |           |
H — C — C = C — C — H        +        H — OH
    |   |   |   |
    H   H   H   H

                    ↓

         H   H   OH  H
         |   |   |   |
     H — C — C — C — C — H
         |   |   |   |
         H   H   H   H
```

A similar reaction can be used to produce 3-methylpentan-3-ol.

Draw a structural formula for the hydrocarbon used to form this molecule. **1**

\+ H—OH
water

↓

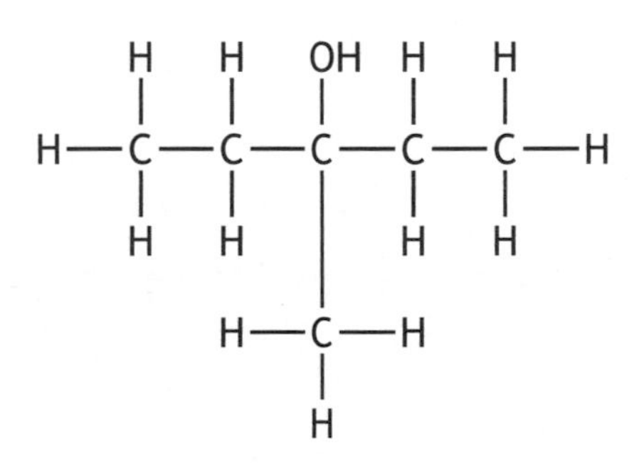

3-methylpentan-3-ol

MARKS | DO NOT WRITE IN THIS MARGIN

13. Succinic acid is a natural antibiotic.

The structure of succinic acid is shown.

```
     O   H   H   O
     ‖   |   |   ‖
HO—C—C—C—C—OH
         |   |
         H   H
```

(a) Name the functional group present in succinic acid. **1**

(b) Succinic acid can form a polymer with ethane-1,2-diol.

The structure of ethane-1,2-diol is shown.

```
          H   H
          |   |
H—O—C—C—O—H
          |   |
          H   H
```

(i) Name the type of polymerisation which would take place between succinic acid and ethane-1,2-diol. **1**

(ii) Draw the repeating unit of the polymer formed between succinic acid and ethane-1,2-diol. **1**

MARKS | DO NOT WRITE IN THIS MARGIN

14. Titanium is the tenth most commonly occurring element in the Earth's crust.

(a) The first step in the extraction of titanium from impure titanium oxide involves the conversion of titanium oxide into titanium(IV) chloride.

$$TiO_2 + 2Cl_2 + 2C \longrightarrow TiCl_4 + 2X$$

(i) Identify **X**. **1**

(ii) Titanium(IV) chloride is a liquid at room temperature and does not conduct electricity.

Suggest the type of bonding that is present in titanium(IV) chloride. **1**

(b) The next step involves separating pure titanium(IV) chloride from other liquid impurities that are also produced during the first step.

Suggest a name for this process. **1**

(c) The equation for the final step in the extraction of titanium is

$$TiCl_4 + 4Na \longrightarrow Ti + 4NaCl$$

The sodium chloride produced can be electrolysed.

Suggest how this could make the extraction of titanium from titanium oxide more economical. **1**

[Turn over

MARKS DO NOT WRITE IN THIS MARGIN

15. Vitamin C is found in fruits and vegetables.

Using iodine solution, a student carried out titrations to determine the concentration of vitamin C in orange juice.

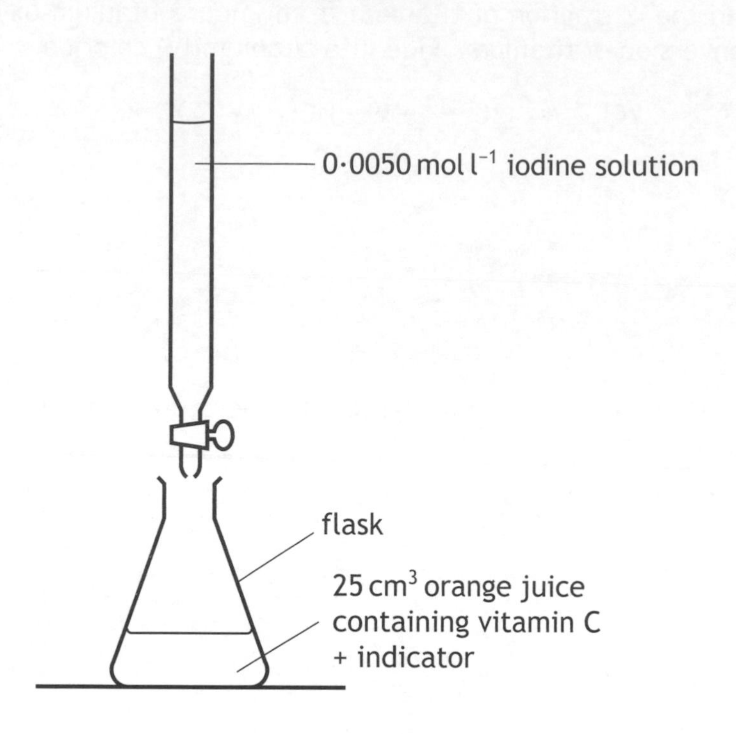

The results of the titration are given in the table.

Titration	*Initial burette reading* (cm³)	*Final burette reading* (cm³)	*Titre* (cm³)
1	1·2	18·0	16·8
2	18·0	33·9	15·9
3	0·5	16·6	16·1

(a) Calculate the average volume, in cm³, that should be used in calculating the concentration of vitamin C. **1**

MARKS | DO NOT WRITE IN THIS MARGIN

15. (continued)

(b) The equation for the reaction is

$$C_6H_8O_6(aq) + I_2(aq) \longrightarrow C_6H_6O_6(aq) + 2HI(aq)$$

vitamin C

Calculate the concentration, in mol l^{-1}, of vitamin C in the orange juice. **3**

Show your working clearly.

[Turn over for Question 16 on *Page twenty-six*

MARKS DO NOT WRITE IN THIS MARGIN

16. A student is given three different compounds each containing carbon.

Using your knowledge of chemistry, describe how the student could identify the compounds. 3

[END OF QUESTION PAPER]

MARKS DO NOT WRITE IN THIS MARGIN

ADDITIONAL SPACE FOR ANSWERS

Question 1(b)

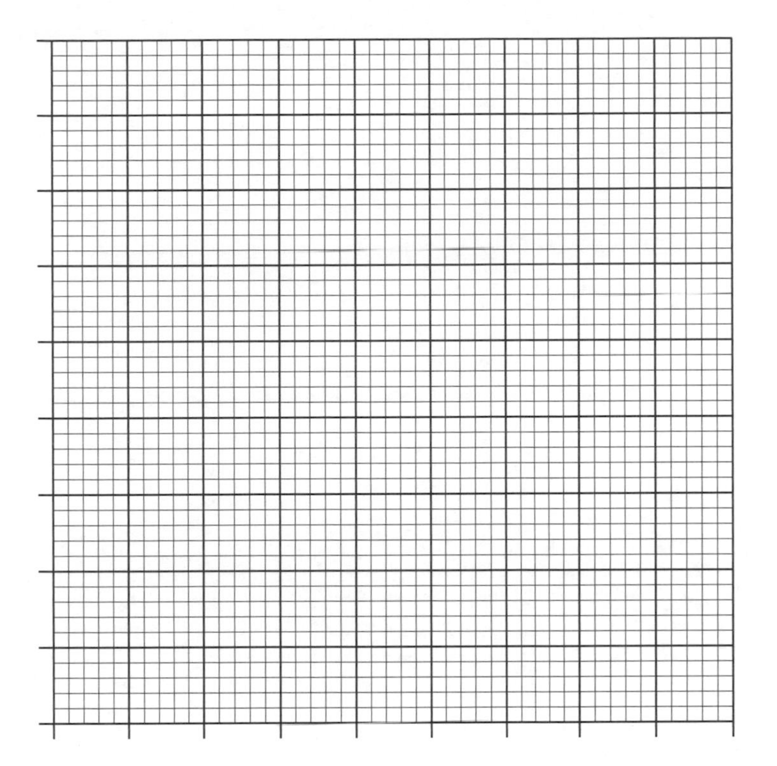

MARKS DO NOT WRITE IN THIS MARGIN

ADDITIONAL SPACE FOR ANSWERS AND ROUGH WORK

MARKS DO NOT WRITE IN THIS MARGIN

ADDITIONAL SPACE FOR ANSWERS AND ROUGH WORK

[BLANK PAGE]

DO NOT WRITE ON THIS PAGE

NATIONAL 5

2016

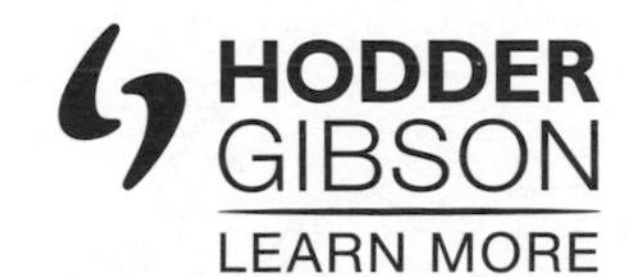

X713/75/02

Chemistry
Section 1 — Questions

WEDNESDAY, 18 MAY

1:00 PM–3:00 PM

Instructions for the completion of Section 1 are given on *Page two* of your question and answer booklet X713/75/01.

Record your answers on the answer grid on *Page three* of your question and answer booklet.

Necessary data will be found in the Chemistry Data Booklet for National 5.

Before leaving the examination room you must give your question and answer booklet to the Invigilator; if you do not, you may lose all the marks for this paper.

SECTION 1

1. When solid sodium chloride dissolves in water, a solution containing sodium ions and chloride ions is formed.

   Which of the following equations correctly shows the state symbols for this process?

   A $NaCl(s) + H_2O(\ell) \longrightarrow Na^+(\ell) + Cl^-(\ell)$

   B $NaCl(s) + H_2O(aq) \longrightarrow Na^+(aq) + Cl^-(aq)$

   C $NaCl(aq) + H_2O(\ell) \longrightarrow Na^+(aq) + Cl^-(aq)$

   D $NaCl(s) + H_2O(\ell) \longrightarrow Na^+(aq) + Cl^-(aq)$

2. The table shows the times taken for 0·5 g of magnesium to react completely with acid under different conditions.

Acid concentration ($mol\ l^{-1}$)	*Temperature* (°C)	*Reaction time* (s)
0·1	20	80
0·1	25	60
0·2	30	20
0·2	40	10

   The time for 0·5 g of magnesium to react completely with 0·2 $mol\ l^{-1}$ acid at 25 °C will be

   A less than 10 s

   B between 10 s and 20 s

   C between 20 s and 60 s

   D more than 80 s.

3. When an atom **X** of an element in Group 1 reacts to become $\mathbf{X}^+$

   A the mass number of **X** decreases

   B the atomic number of **X** increases

   C the charge of the nucleus increases

   D the number of occupied energy levels decreases.

4. Which of the following does **not** contain covalent bonds?

   A Sulfur

   B Copper

   C Oxygen

   D Hydrogen

5. Which of the following structures is **never** found in compounds?

A Ionic

B Monatomic

C Covalent network

D Covalent molecular

6. Which line in the table shows the properties of an ionic substance?

			Conducts electricity	
	Melting point (°C)	*Boiling point* (°C)	*Solid*	*Liquid*
A	19	80	no	no
B	655	1425	no	no
C	1450	1740	no	yes
D	1495	2927	yes	yes

7. What is the name of the compound with the formula Ag_2O?

A Silver(I) oxide

B Silver(II) oxide

C Silver(III) oxide

D Silver(IV) oxide

8. An element was burned in air. The product was added to water, producing a solution with a pH less than 7. The element could be

A tin

B zinc

C sulfur

D sodium.

9. When methane burns in a plentiful supply of air, the products are

A carbon and water

B carbon dioxide and water

C carbon monoxide and water

D carbon dioxide and hydrogen.

[Turn over

10. Which of the following compounds belongs to the same homologous series as the compound with the molecular formula C_3H_8?

A

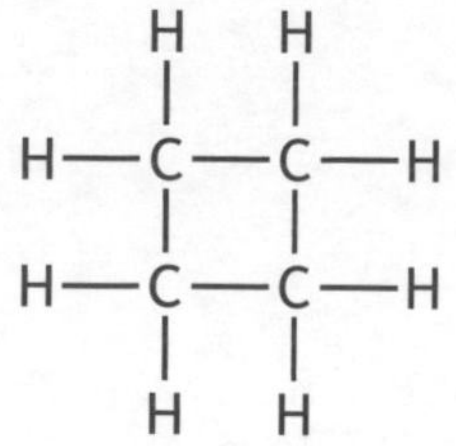

```
     H   H
     |   |
  H—C———C—H
     |   |
  H—C———C—H
     |   |
     H   H
```

B

```
     H           H
     |           |
  H—C———C═C———C—H
     |   |   |   |
     H   H   H   H
```

C

```
             H
             |
          H—C—H
             |
     H   H   |   H
     |   |   |   |
  H—C———C———C———C—H
     |   |   |   |
     H   H   H   H
```

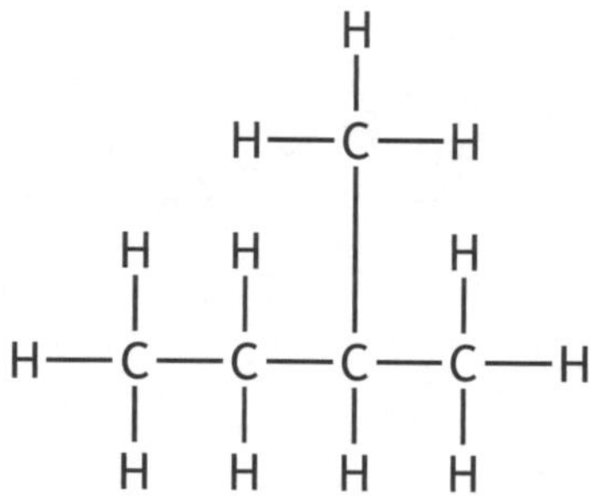

D

```
             H
             |
          H—C—H
             |
     H   H   |
     |   |   |
  H—C———C———C═C—H
     |   |       |
     H   H       H
```

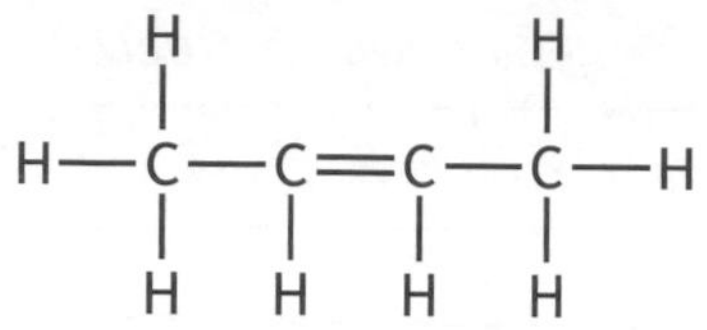

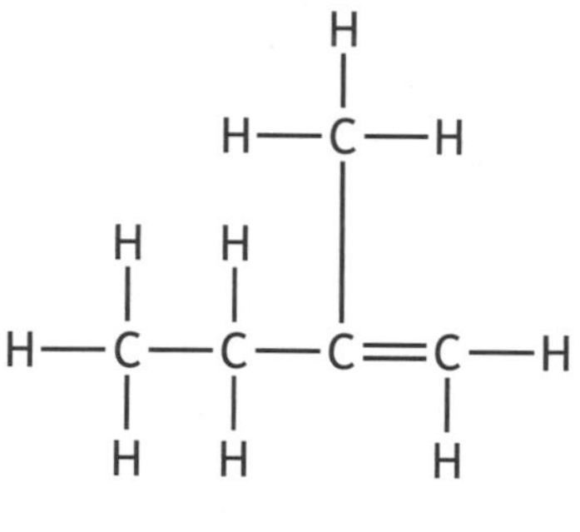

11.

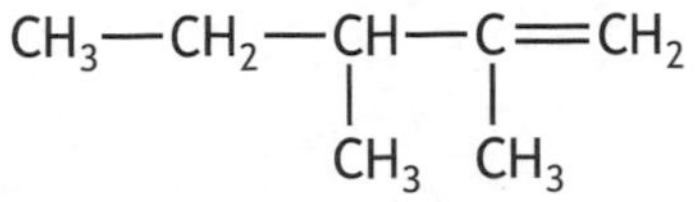

```
CH3—CH2—CH—C═CH2
         |  |
        CH3 CH3
```

The systematic name for the structure shown is

A 1,2-dimethylpent-1-ene

B 2,3-dimethylpent-1-ene

C 3,4-dimethylpent-4-ene

D 3,4-dimethylpent-1-ene.

12. Two isomers of butene are

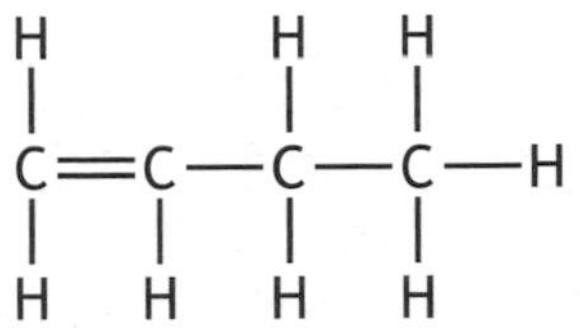

```
H     H  H                 H           H
|     |  |                 |           |
C═C—C—C—H      and     H—C—C═C—C—H
| | | |                    |  |  |  |
H H H H                    H  H  H  H
```

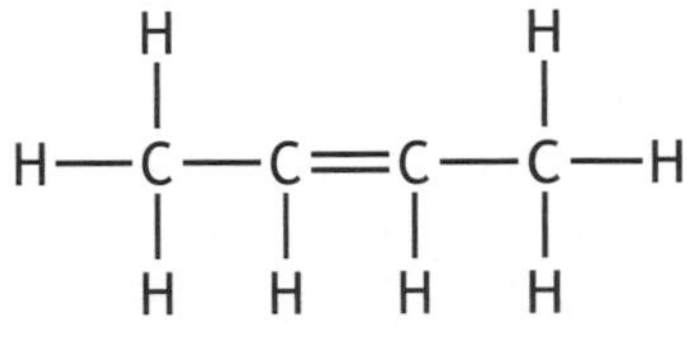

Which of the following structures represents a third isomer of butene?

A

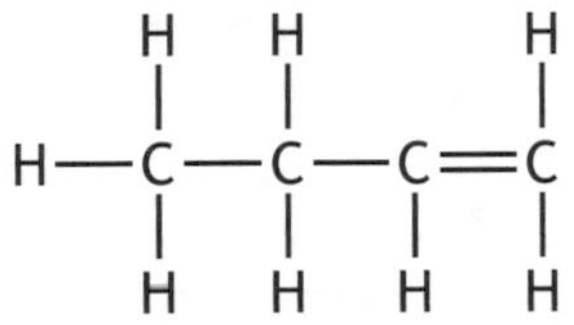

```
    H  H     H
    |  |     |
H—C—C—C═C
    |  |  |  |
    H  H  H  H
```

B

```
       H
       |
    H—C—H
       |
    H  |  H
    |  |  |
H—C—C═C
    |     |
    H     H
```

C

```
    H  H
    |  |
H—C—C—H
    |  |
H—C═C—H
```

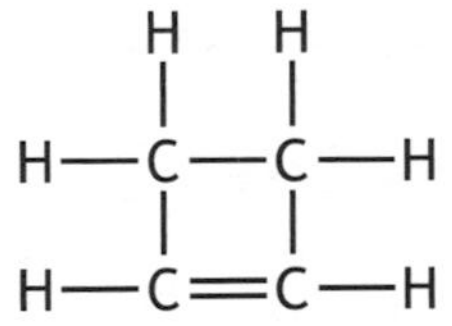

D

```
H     H     H
|     |     |
C═C—C—C═C
| | | | |
H H H H H
```

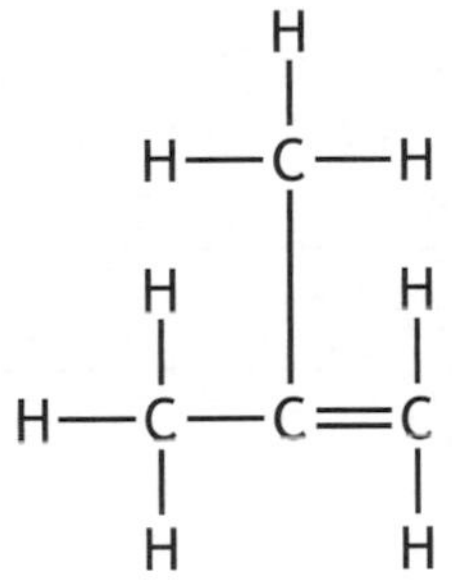

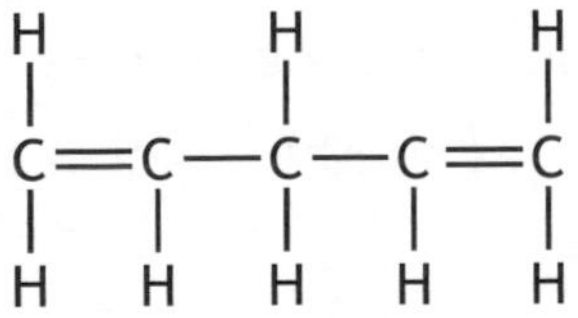

[Turn over

13. Which of the following structures represents an ester?

A

```
     H   O        H
     |   ||       |
H ── C ── C ── O ── C ── H
     |            |
     H            H
```

B

```
     H   O
     |   ||
H ── C ── C ── O ── H
     |
     H
```

C

```
     H   O
     |   ||
H ── C ── C ── H
     |
     H
```

D

```
     H   O   H
     |   ||  |
H ── C ── C ── C ── H
     |       |
     H       H
```

14. The lowest temperature at which a hydrocarbon ignites is called its flash point.

Hydrocarbon	*Formula*	*Boiling point* (°C)	*Flash point* (°C)
hexene	C_6H_{12}	63	−25
hexane	C_6H_{14}	69	−23
cyclohexane	C_6H_{12}	81	−20
heptane	C_7H_{16}	98	−1
octane	C_8H_{18}	126	15

Using information in the table, identify the correct statement.

A Octane will ignite at 0 °C.

B Hydrocarbons with the same molecular mass have the same flash point.

C The flash point of a hydrocarbon increases as the boiling point increases.

D In a homologous series the flash point decreases as the number of carbon atoms increases.

15. Which of the following metals can be obtained from its ore by heating with carbon monoxide?

You may wish to use the data booklet to help you.

A Magnesium

B Aluminium

C Calcium

D Nickel

16. Polyesters are always made from monomers

A which are the same

B which are unsaturated

C with one functional group per molecule

D with two functional groups per molecule.

17. Some smoke detectors make use of radiation which is very easily stopped by tiny smoke particles moving between the radioactive source and the detector.

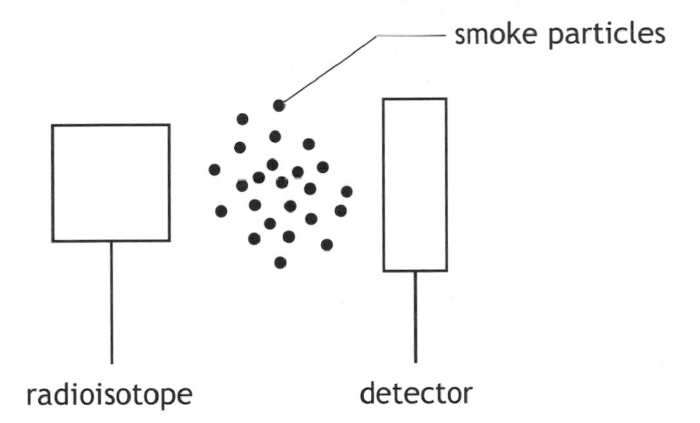

The most suitable type of radioisotope for a smoke detector would be

A an alpha-emitter with a long half-life

B a gamma-emitter with a short half-life

C an alpha-emitter with a short half-life

D a gamma-emitter with a long half-life.

[Turn over for next question

18. Which particle will be formed when an atom of $^{234}_{90}Th$ emits a β-particle?

A $^{234}_{91}Pa$

B $^{230}_{88}Ra$

C $^{234}_{89}Ac$

D $^{238}_{92}U$

19. ^{14}C has a half life of 5600 years. An analysis of charcoal from a wood fire shows that its ^{14}C content is 25% of that in living wood.

How many years have passed since the wood for the fire was cut?

A 1400

B 4200

C 11 200

D 16 800

20. A solution of potassium carbonate, made up using tap water, was found to be cloudy. This could result from the tap water containing

A lithium ions

B calcium ions

C sodium ions

D ammonium ions.

You may wish to use the data booklet to help you.

[END OF SECTION 1. NOW ATTEMPT THE QUESTIONS IN SECTION 2 OF YOUR QUESTION AND ANSWER BOOKLET]

FOR OFFICIAL USE

National Qualifications 2016

Mark

X713/75/01

Chemistry Section 1 — Answer Grid And Section 2

WEDNESDAY, 18 MAY

1:00 PM – 3:00 PM

Fill in these boxes and read what is printed below.

Full name of centre

Town

Forename(s)

Surname

Number of seat

Date of birth

Day Month Year

Scottish candidate number

Total marks — 80

SECTION 1 — 20 marks

Attempt ALL questions.

Instructions for the completion of Section 1 are given on *Page two*.

SECTION 2 — 60 marks

Attempt ALL questions.

Necessary Data will be found in the Chemistry Data Booklet for National 5.

Write your answers clearly in the spaces provided in this booklet. Additional space for answers and rough work is provided at the end of this booklet. If you use this space you must clearly identify the question number you are attempting. Any rough work must be written in this booklet. You should score through your rough work when you have written your final copy.

Use **blue** or **black** ink.

Before leaving the examination room you must give this booklet to the Invigilator; if you do not, you may lose all the marks for this paper.

SECTION 1 — 20 marks

The questions for Section 1 are contained in the question paper X713/75/02.

Read these and record your answers on the answer grid on *Page three* opposite.

Use **blue** or **black** ink. Do NOT use gel pens or pencil.

1. The answer to each question is **either** A, B, C or D. Decide what your answer is, then fill in the appropriate bubble (see sample question below).

2. There is **only one correct** answer to each question.

3. Any rough working should be done on the additional space for answers and rough work at the end of this booklet.

Sample Question

To show that the ink in a ball-pen consists of a mixture of dyes, the method of separation would be

A fractional distillation

B chromatography

C fractional crystallisation

D filtration.

The correct answer is **B**—chromatography. The answer **B** bubble has been clearly filled in (see below).

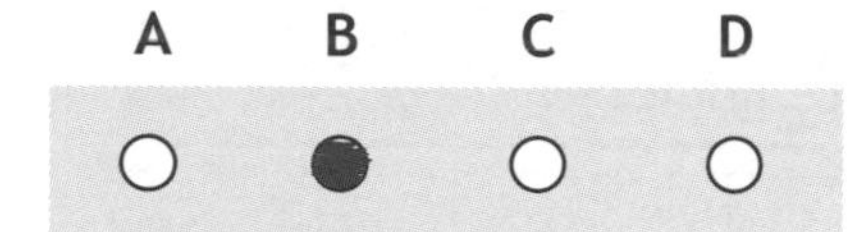

Changing an answer

If you decide to change your answer, cancel your first answer by putting a cross through it (see below) and fill in the answer you want. The answer below has been changed to **D**.

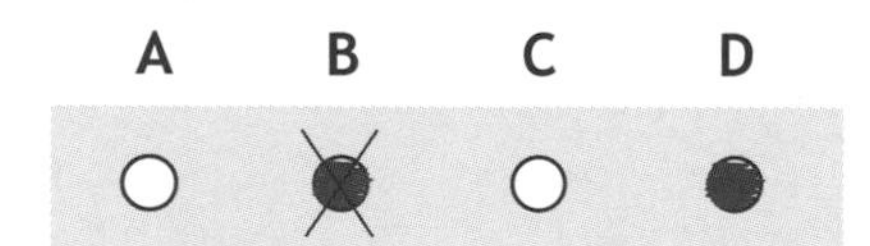

If you then decide to change back to an answer you have already scored out, put a tick (✓) to the **right** of the answer you want, as shown below:

A B C D

○ ⊗✓ ○ ⊗

or

A B C D

○ ⊗✓ ○ ○

SECTION 1 — Answer Grid

	A	B	C	D
1	○	○	○	○
2	○	○	○	○
3	○	○	○	○
4	○	○	○	○
5	○	○	○	○
6	○	○	○	○
7	○	○	○	○
8	○	○	○	○
9	○	○	○	○
10	○	○	○	○
11	○	○	○	○
12	○	○	○	○
13	○	○	○	○
14	○	○	○	○
15	○	○	○	○
16	○	○	○	○
17	○	○	○	○
18	○	○	○	○
19	○	○	○	○
20	○	○	○	○

[Turn over

[BLANK PAGE]

DO NOT WRITE ON THIS PAGE

[Turn over for next question

DO NOT WRITE ON THIS PAGE

MARKS | DO NOT WRITE IN THIS MARGIN

SECTION 2 — 60 marks

Attempt ALL questions

1. Elements are made up of atoms.

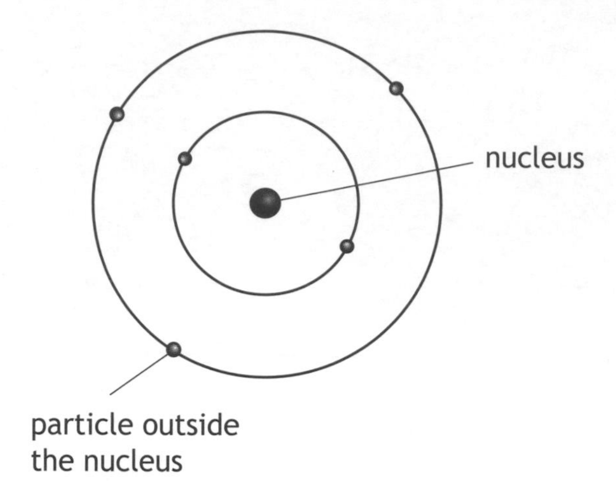

(a) Complete the tables to show the missing information.

(i) **1**

In the Nucleus		
Particle	*Relative Mass*	*Charge*
proton		+1
neutron	1	

(ii) **1**

Outside the Nucleus		
Particle	*Relative Mass*	*Charge*
	almost zero	

(b) A sample of nitrogen was found to contain equal amounts of two isotopes. One isotope has mass number 14 and the other has mass number 15.

What is the relative atomic mass of this sample of nitrogen? **1**

MARKS | DO NOT WRITE IN THIS MARGIN

1. (continued)

(c) Nitrogen can form bonds with other elements.

The diagram shows the shape of a molecule of ammonia (NH_3).

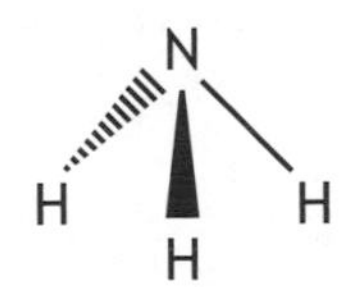

(i) State the name used to describe the shape of a molecule of ammonia. **1**

(ii) Name the industrial process used to manufacture ammonia. **1**

[Turn over

MARKS | DO NOT WRITE IN THIS MARGIN

2. The monomer used to produce polystyrene has the following structure.

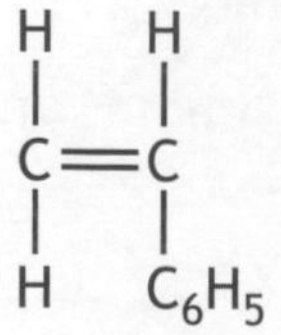

```
H   H
|   |
C = C
|   |
H   C6H5
```

styrene

(a) Draw a section of polystyrene, showing three monomer units joined together. **1**

(b) When two different monomers polymerise, a copolymer is formed as shown.

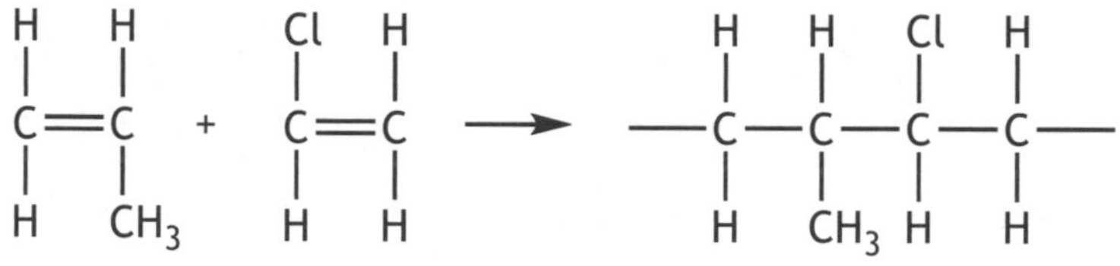

```
H   H          Cl  H                  H   H    Cl  H
|   |          |   |                  |   |    |   |
C = C     +    C = C      ──►      ── C ─ C ── C ─ C ──
|   |          |   |                  |   |    |   |
H   CH3        H   H                  H   CH3  H   H
```

Another copolymer can be made from styrene and acrylonitrile monomers. A section of this copolymer is shown below.

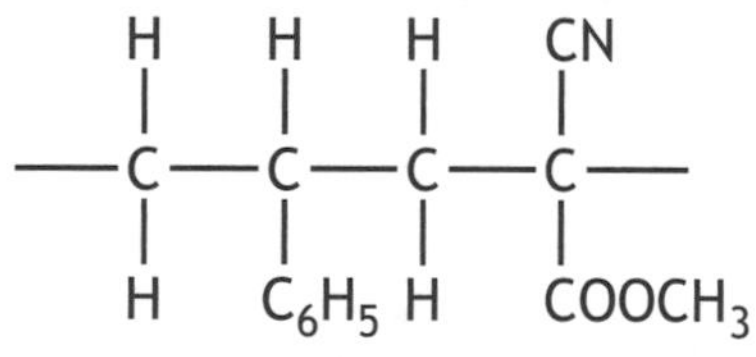

```
     H    H     H    CN
     |    |     |    |
  ── C ── C ─── C ── C ──
     |    |     |    |
     H    C6H5  H    COOCH3
```

Draw the structure of the acrylonitrile monomer. **1**

[Turn over for next question

DO NOT WRITE ON THIS PAGE

MARKS | DO NOT WRITE IN THIS MARGIN

3. Hydrogen gas can be produced in the laboratory by adding a metal to dilute acid. Heat energy is also produced in the reaction.

(a) State the term used to describe all chemical reactions that release heat energy. **1**

(b) A student measured the volume of hydrogen gas produced when zinc lumps were added to dilute hydrochloric acid.

Time (s)	0	10	20	30	40	50	60	70
Volume of hydrogen (cm^3)	0	12	21	29	34	36	37	37

(i) Calculate the average rate of reaction, in $cm^3 s^{-1}$, between 10 and 30 seconds. **2**

Show your working clearly.

(ii) Estimate the time taken, in seconds, for the reaction to finish. **1**

(iii) The student repeated the experiment using the same mass of zinc.

State the effect on the rate of the reaction if zinc powder was used instead of lumps. **1**

MARKS | DO NOT WRITE IN THIS MARGIN

3. (continued)

(c) Another student reacted aluminium with dilute nitric acid.

$$2Al(s) + 6HNO_3(aq) \longrightarrow 2Al(NO_3)_3(aq) + 3H_2(g)$$

(i) Circle the formula for the salt in the above equation. **1**

(ii) 1 mole of hydrogen gas has a volume of 24 litres.

Calculate the volume of hydrogen gas, in litres, produced when 0·01 moles of aluminium react with dilute nitric acid. **2**

Show your working clearly.

[Turn over

MARKS DO NOT WRITE IN THIS MARGIN

4. Some rocks contain the mineral with the formula Al_2SiO_5.

This mineral exists in three different forms, andalusite, sillimanite, and kyanite. The form depends on the temperature and pressure.

The diagram shows this relationship.

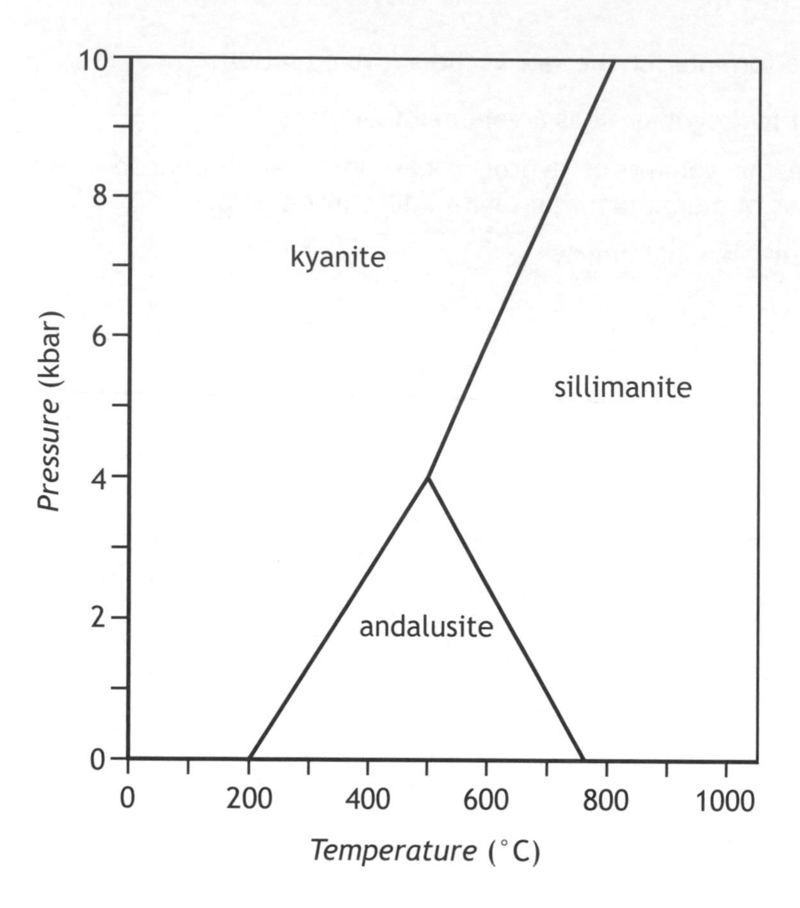

(a) (i) Name the **two** forms which could exist at 400 °C. **1**

(ii) Complete the table to show the temperature and pressure at which all three forms would exist. **1**

Temperature (°C)	
Pressure (kbar)	

MARKS DO NOT WRITE IN THIS MARGIN

4. (continued)

(b) Calculate the percentage mass of silicon in andalusite, Al_2SiO_5. 3

Show your working clearly.

[Turn over

MARKS | DO NOT WRITE IN THIS MARGIN

5. Read the passage and answer the questions that follow.

Gold — a very useful metal

Gold has been associated with wealth since before the first gold coins were minted in Lydia (modern Turkey) about 550 BC. It does not react with water, air, alkalis and almost all acids. Gold only has one naturally occurring isotope with mass 197.

As an element it has many uses in the modern world. 1 gram of gold can be beaten into a gold film covering one square metre and thin coatings of gold are used as lubricants in aerospace applications. Gold electroplating can be used to coat electrical connectors and printed circuit boards.

Chemists have recently discovered that gold nanoparticles make superb catalysts for many reactions such as the conversion of alcohols into aldehydes and ketones. It can also be used as a catalyst for removing trace carbon monoxide from gases. In this reaction carbon monoxide reacts with oxygen to form carbon dioxide.

Gold nanorods can be grown from a dilute solution of auric acid and are used in the treatment of some forms of cancer.

Adapted from *Education in Chemistry*, Volume 45, November 2008

(a) Suggest a reason why gold was used in the first coins minted. **1**

(b) Calculate the number of neutrons present in the naturally occurring isotope of gold. **1**

You may wish to use the data booklet to help you.

MARKS DO NOT WRITE IN THIS MARGIN

5. (continued)

(c) (i) Write an equation, using symbols and formulae, to show the reaction for removing trace carbon monoxide from gases.

There is no need to balance this equation. **1**

(ii) State the role of gold in this reaction. **1**

(d) Circle the correct words to complete the sentence. **1**

Gold nanorods can be grown from a solution which contains

more {hydroxide / hydrogen} ions than {hydroxide / hydrogen} ions.

[Turn over

MARKS | DO NOT WRITE IN THIS MARGIN

6. (a) A fertiliser for tomato plants contains compounds of phosphorus and potassium.

(i) Suggest an experimental test, including the result, to show that potassium is present in the fertiliser. **1**

You may wish to use the data booklet to help you.

(ii) Ammonium citrate is included in the fertiliser because some phosphorus compounds are more soluble in ammonium citrate solution than they are in water.

Suggest another reason why ammonium citrate is added to the fertiliser. **1**

(b) In the production of the fertiliser ammonium phosphate, phosphoric acid (H_3PO_4) reacts with ammonium hydroxide as shown.

$$H_3PO_4(aq) \quad + \quad NH_4OH(aq) \longrightarrow (NH_4)_3PO_4(aq) \quad + \quad H_2O(\ell)$$

Balance this equation. **1**

MARKS | DO NOT WRITE IN THIS MARGIN

7. The element strontium was discovered in 1790 in the village of Strontian in Scotland.

Using your knowledge of chemistry, comment on the chemistry of strontium. 3

[Turn over

MARKS | DO NOT WRITE IN THIS MARGIN

8. Essential oils can be extracted from plants and used in perfumes and food flavourings.

(a) Essential oils contain compounds called terpenes.

A terpene is a chemical made up of a number of isoprene molecules joined together.

The shortened structural formula of isoprene is $CH_2C(CH_3)CHCH_2$.

Draw the full structural formula for isoprene. **1**

(b) Essential oils can be extracted from the zest of lemons in the laboratory by steam distillation.

The process involves heating up water in a boiling tube until it boils. The steam produced then passes over the lemon zest which is separated from the water by glass wool. As the steam passes over the lemon zest it carries the essential oils into a delivery tube. The condensed liquids (essential oils and water) are collected in a test tube placed in a cold water bath.

Complete the diagram to show the apparatus required to collect the essential oils. **1**

(An additional diagram, if required, can be found on *Page twenty-nine*.)

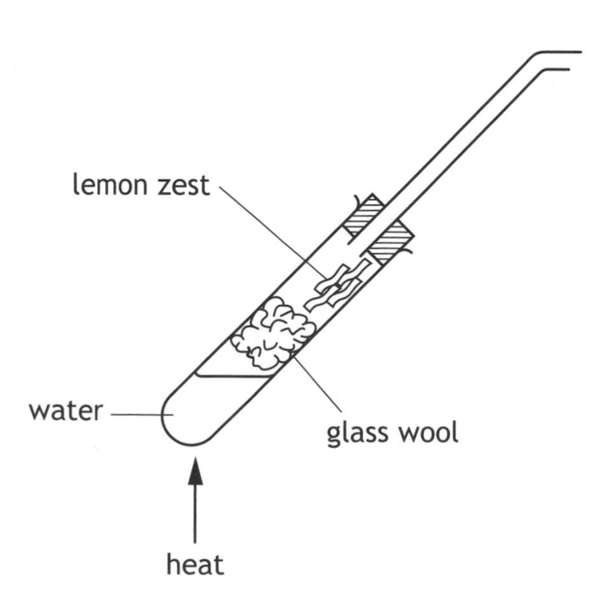

MARKS | DO NOT WRITE IN THIS MARGIN

8. (continued)

(c) Limonene, $C_{10}H_{16}$, is an essential oil which is added to some cleaning products to give them a lemon scent.

CH_3 — C (=CH) ring: H_2C, H_2C, CH, CH_2; CH — C (H_3C, $=CH_2$)

The concentration of limonene present in a cleaning product can be determined by titrating with bromine solution.

(i) Name the type of chemical reaction taking place when limonene reacts with bromine solution. **1**

(ii) Write the molecular formula for the product formed when limonene, $C_{10}H_{16}$, reacts completely with bromine solution. **1**

[Turn over

MARKS | DO NOT WRITE IN THIS MARGIN

9. Ethanol can be used as an alternative fuel for cars.

(a) A student considered two methods to confirm the amount of energy released when ethanol burns.

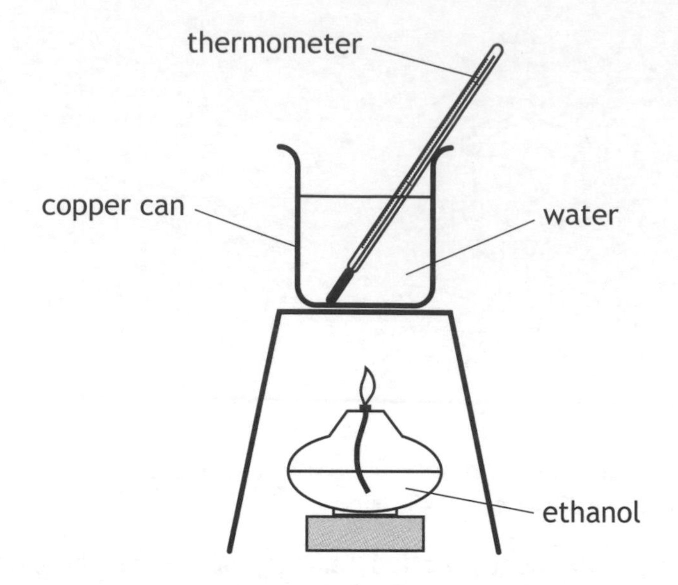

Method A	*Method B*
1. Record the initial temperature of the water.	1. Record the initial temperature of the water.
2. Weigh the burner containing the fuel.	2. Weigh the burner containing the fuel.
3. Place the burner under the copper can and then light the burner.	3. Light the burner and then place it under the copper can.
4. Extinguish the flame after 2 minutes.	4. Extinguish the flame after 2 minutes.
5. Record the final temperature and reweigh the burner.	5. Record the final temperature and reweigh the burner.

Explain which method would give a more accurate result. **2**

MARKS | DO NOT WRITE IN THIS MARGIN

9. (continued)

(b) The table gives information about the amount of energy released when 1 mole of some alcohols are burned.

Name of alcohol	*Energy released when one mole of alcohol is burned* (kJ)
propan-1-ol	2021
propan-2-ol	2005
butan-1-ol	2676
butan-2-ol	2661
pentan-1-ol	3329
pentan-2-ol	3315
hexan-1-ol	3984

(i) Write a statement linking the amount of energy released to the position of the functional group in an alcohol molecule. **1**

(ii) Predict the amount of energy released, in kJ, when 1 mole of hexan-2-ol is burned. **1**

(c) Ethanol can also be used in portable camping stoves.

The chemical reaction in a camping stove releases 23 kJ of energy. If 100 g of water is heated using this stove, calculate the rise in temperature of the water, in °C. **3**

You may wish to use the data booklet to help you.

Show your working clearly.

[Turn over

MARKS | DO NOT WRITE IN THIS MARGIN

10. A battery is a number of cells joined together.

(a) The diagram shows a simple battery made from copper and zinc discs separated by paper soaked in potassium nitrate solution.

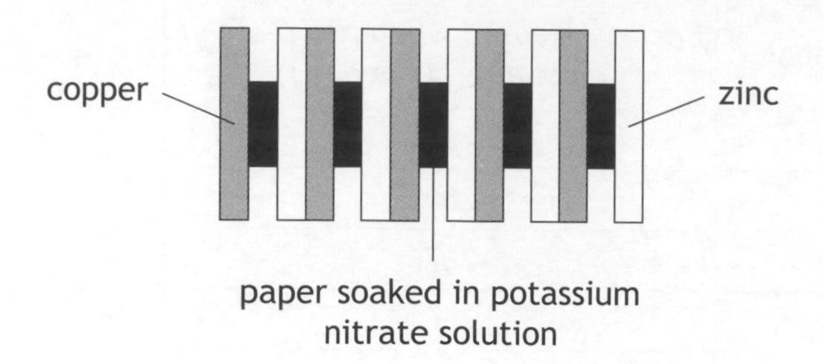

The purpose of the potassium nitrate solution is to complete the circuit.

State the **term** used to describe an ionic compound which is used for this purpose. **1**

(b) A student set up a cell using the same metals as those used in the battery.

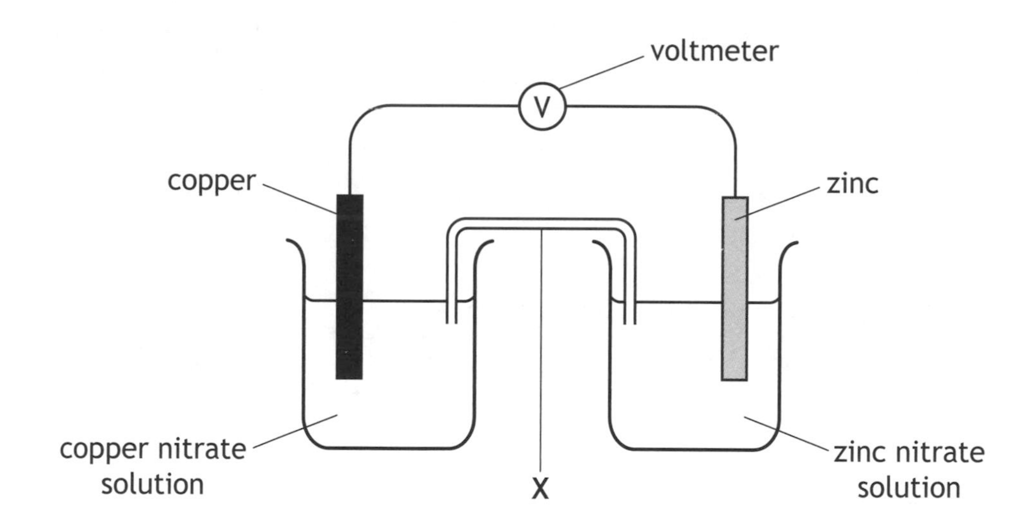

(i) **On the diagram**, draw an arrow to show the path and direction of electron flow. **1**

You may wish to use the data booklet to help you.

(ii) Name the piece of apparatus labelled **X**. **1**

MARKS | DO NOT WRITE IN THIS MARGIN

10. (continued)

(c) Electricity can also be produced in a cell containing non-metals.

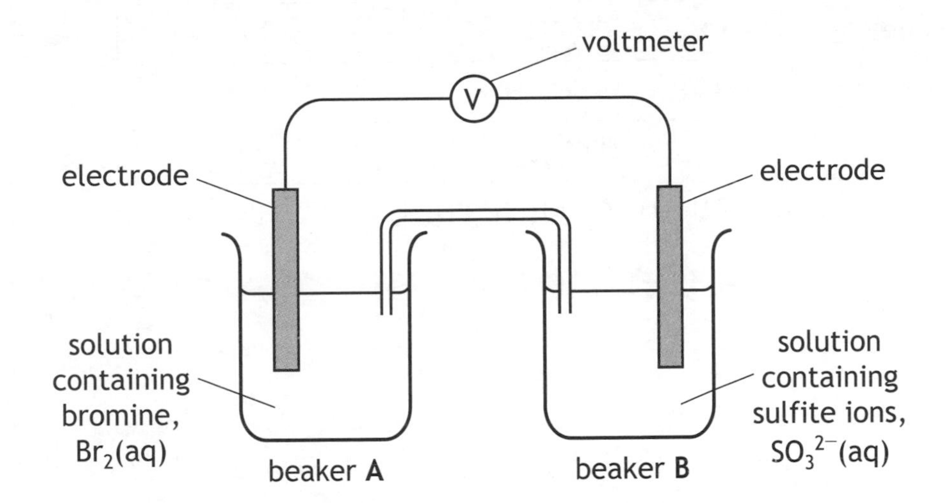

The reactions occurring at each electrode are

Beaker **A** $Br_2(\ell) + 2e^- \longrightarrow 2Br^-(aq)$

Beaker **B** $SO_3^{2-}(aq) + H_2O(\ell) \longrightarrow SO_4^{2-}(aq) + 2H^+(aq) + 2e^-$

(i) Name the type of chemical reaction taking place in beaker **B**. **1**

(ii) Write the redox equation for the overall reaction. **1**

(iii) Name a non-metal element which is suitable for use as the electrodes. **1**

[Turn over

MARKS DO NOT WRITE IN THIS MARGIN

11. Ethers are a group of compounds containing carbon, hydrogen and oxygen.

Name of ether	*Structural formula*	*Boiling point* (°C)
methoxyethane	$CH_3-O-CH_2CH_3$	7
ethoxyethane	$CH_3CH_2-O-CH_2CH_3$	35
X	$CH_3-O-CH_2CH_2CH_3$	39
propoxybutane	$CH_3CH_2CH_2-O-CH_2CH_2CH_2CH_3$	117

(a) Name ether **X**. **1**

(b) Suggest a general formula for this homologous series. **1**

(c) Methoxyethane is a covalent molecular substance. It has a low boiling point and is a gas at room temperature.

Circle the correct words to complete the sentence. **1**

The bonds between the molecules are {weak / strong} and the bonds within the molecule are {weak / strong}.

MARKS | DO NOT WRITE IN THIS MARGIN

11. (continued)

(d) Epoxides are a family of cyclic ethers.

The full structural formula for the first member of this family is shown.

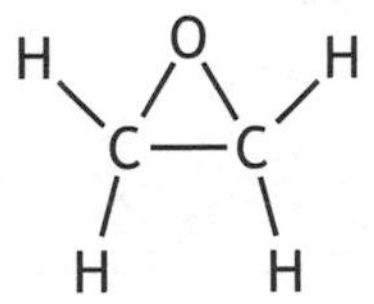

(i) Epoxides can be produced by reacting an alkene with oxygen.

Name the alkene which would be used to produce the epoxide shown. **1**

(ii) Epoxides have three atoms in a ring, one of which is oxygen.

Draw a structural formula for the epoxide with the chemical formula C_3H_6O. **1**

[Turn over

MARKS | DO NOT WRITE IN THIS MARGIN

12. Betanin is responsible for the red colour in beetroot and can be used as a food colouring.

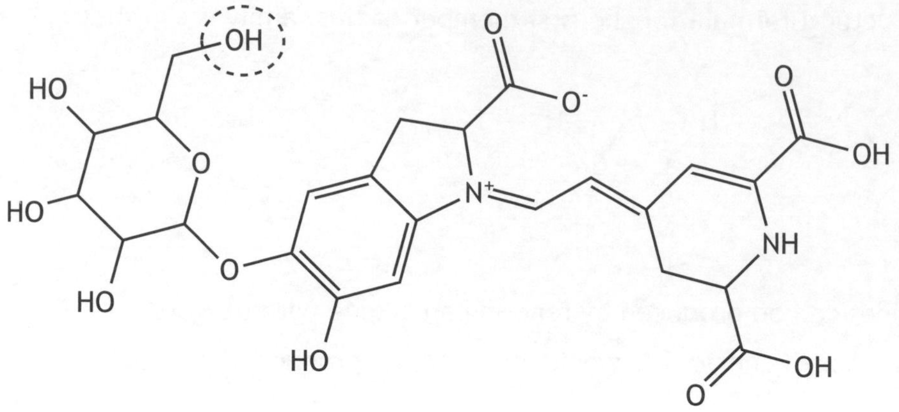

(a) Name the functional group circled in the diagram above. **1**

(b) Betanin can be used as an indicator in a neutralisation reaction.

The pH range at which some indicators change colour is shown.

Indicator	*pH range of colour change*
methyl orange	3·2 to 4·4
litmus	5·0 to 8·0
phenolphthalein	8·2 to 10·0
betanin	9·0 to 10·0

The indicator used in a neutralisation reaction depends on the pH at the end point.

The table below shows the end point of neutralisation reactions using different types of acid and base.

Type of acid	*Type of base*	*pH at the end point*
strong	strong	7
strong	weak	below 7
weak	strong	above 7

Betanin can be used to indicate the end point in the reaction between oxalic acid and sodium hydroxide solution.

State the type of acid **and** the type of base used in this reaction. **1**

12. (continued) MARKS DO NOT WRITE IN THIS MARGIN

(c) A student carried out a titration experiment to determine the concentration of a sodium hydroxide solution.

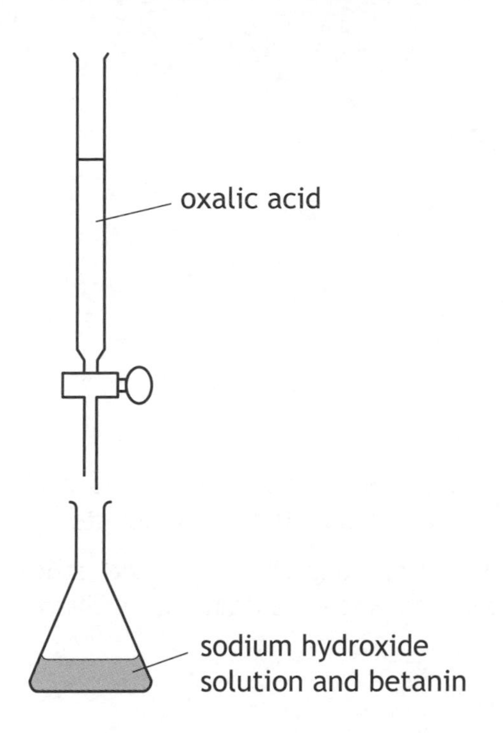

	Initial burette reading (cm^3)	*Final burette reading* (cm^3)	*Volume used* (cm^3)
Rough titre	0·0	15·6	15·6
1st titre	15·6	30·5	14·9
2nd titre	30·5	45·6	15·1

Using the results in the table, calculate the average volume, in cm^3, of oxalic acid required to neutralise the sodium hydroxide solution. **1**

(d) Oxalic acid is found naturally in rhubarb. A piece of rhubarb was found to contain 1·8 g of oxalic acid.

Calculate the number of moles of oxalic acid contained in the piece of rhubarb. **1**

(Formula mass of oxalic acid = 90)

[Turn over for next question

MARKS | DO NOT WRITE IN THIS MARGIN

13. Carbonated water, also known as sparkling water, is water into which carbon dioxide gas has been dissolved. This process is called carbonating.

A group of students are given two brands of carbonated water and asked to determine which brand contains more dissolved carbon dioxide.

Using your knowledge of chemistry, describe how the students could determine which brand of carbonated water contains more dissolved carbon dioxide. **3**

[END OF QUESTION PAPER]

MARKS DO NOT WRITE IN THIS MARGIN

ADDITIONAL SPACE FOR ANSWERS

Additional diagram for Question 8 (b)

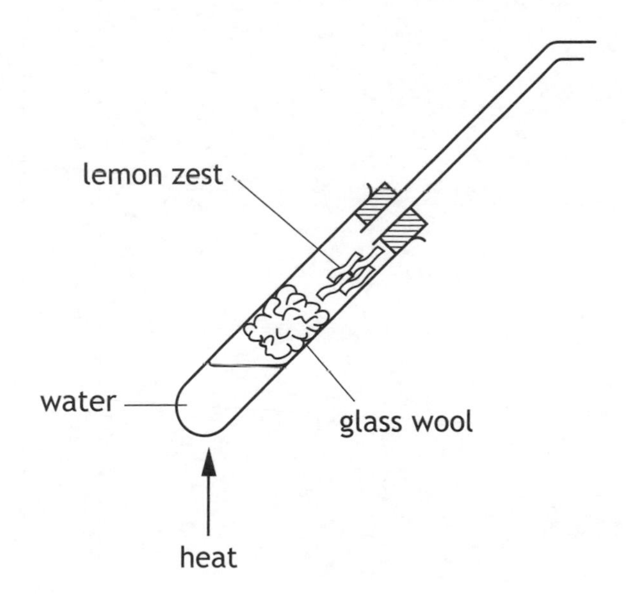

MARKS | DO NOT WRITE IN THIS MARGIN

ADDITIONAL SPACE FOR ANSWERS AND ROUGH WORK

MARKS DO NOT WRITE IN THIS MARGIN

ADDITIONAL SPACE FOR ANSWERS AND ROUGH WORK

[BLANK PAGE]

DO NOT WRITE ON THIS PAGE

NATIONAL 5

Answers

ANSWERS FOR

SQA NATIONAL 5 CHEMISTRY 2016

NATIONAL 5 CHEMISTRY 2014

Section 1

Question	Response
1.	A
2.	D
3.	A
4.	C
5.	C
6.	A
7.	D
8.	A
9.	C
10.	B
11.	C
12.	B
13.	C
14.	C
15.	A
16.	D
17.	B
18.	B
19.	B
20.	D

Section 2

1. (a) **Repulsion/repelled** by nucleus/positive nucleus /protons/positive protons/positive particles in nucleus or in atom or in gold/ like charges in nucleus, atom or gold

(b) (i) Protons – 79
Electrons – 79
Neutrons – 118

(ii) Same atomic number/protons AND different mass number/mass/number of neutrons

Atoms of the same element with different mass number/mass/number of neutrons

2. (a) Covalent network
Ionic lattice
Metallic lattice
(Discrete) covalent molecular

(b) Delocalised /able or free to move

3. (a) Potassium is an essential element
or
humans/human body cannot store it/have no mechanism for storing it

(b) 0·022 or 0·02

(c) Lilac/purple

(d) K^+ NO_3^-

4. (a)

```
   NC12H8      H   NC12H8      H   NC12H8      H
     |         |     |         |     |         |
  —  C  ———————C  —  C  ———————C  —  C  ———————C  —
     |         |     |         |     |         |
     H         H     H         H     H         H
```

(b) Addition

5. (a) alpha or α

(b) ¼/0·25/25%

(c) Sodium/Na
$^{24}_{11}Na$ ^{24}Na $_{11}Na$

6. This is an open ended question
1 mark: The student has demonstrated a limited understanding of the chemistry involved. The candidate has made some statement(s) which is/are relevant to the situation, showing that at least a little of the chemistry within the problem is understood.
2 marks: The student has demonstrated a reasonable understanding of the chemistry involved. The student makes some statement(s) which is/are relevant to the situation, showing that the problem is understood.
3 marks: The maximum available mark would be awarded to a student who has demonstrated a good understanding of the chemistry involved. The student shows a good comprehension of the chemistry of the situation and has provided a logically correct answer to the question posed. This type of response might include a statement of the principles involved, a relationship or an equation, and the application of these to respond to the problem. This does not mean the answer has to be what might be termed an "excellent" answer or a "complete" one.

7. (a) (i) Haber

(ii) Diagram showing three hydrogen atoms and one nitrogen atom with three pairs of bonding electrons and two non-bonding electrons in nitrogen eg

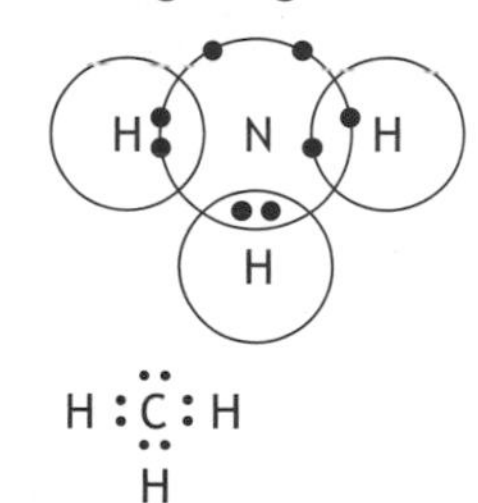


```
   ••
H : C : H
   ••
   H
```

(b) (i) Water/H_2O/Hydrogen oxide

(ii) Arrow from nitrogen monoxide from absorber to nitrogen monoxide below reactor (anywhere below the reactor and above nitrogen dioxide)

(c) (i) Neutralisation

(ii) Evaporation
or
boil it/boil off the water
or
distillation

8. (a) Perfumes, solvents, flavourings, fragrances, preservatives
   (b) (i) Hydroxyl
   (ii) Any correct full or shortened structural formula for an isomer
   (iii) $C_nH_{2n}O_2$
   $C_nH_{2n+1}COOH$
   (c) ethanol
   propanoic acid

9. (a) They have similar chemical properties
   and
   They have the same general formula.
   (b) Butane, or it, has stronger/more/bigger forces of attraction between molecules or mention of intermolecular attractions
   (c) 2090
   (d) Produces SO_2/acidic gases/oxides of sulfur
   Produces acid rain

10. (a) (i) The higher/lower the number of carbon atoms the higher/lower the flash point
   The flash point increases/decreases as the number of carbon atoms increases/decreases
   (ii) 47 – 51 inclusive
   (b) 99

11. (a) $2Cl^- \longrightarrow Cl_2 + 2e^-$
   or
   $2Cl^- - 2e^- \longrightarrow Cl_2$
   (b) (i) sodium hydroxide
   or
   sodium oxide
   (ii) Water is the **only** product
   Hydrogen is infinite/renewable
   Doesn't produce greenhouse gases/CO_2/CO
   The products/gases produced do not contribute to the greenhouse effect/global warming
   Fossil fuels not being used up as fuel
   (c)

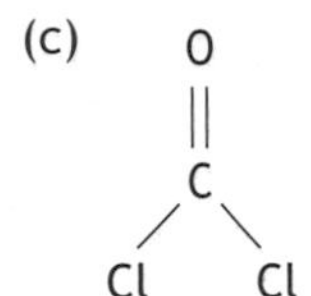

12. (a) Reduction
   (b) 70
   (c) Electrolysis

13. (a) 16
   (b) 0·08

14. This is an open-ended question.
1 mark: The student has demonstrated a limited understanding of the chemistry involved. The candidate has made some statement(s) which is/are relevant to the situation, showing that at least a little of the chemistry within the problem is understood.
2 marks: The student has demonstrated a reasonable understanding of the chemistry involved. The student makes some statement(s) which is/are relevant to the situation, showing that the problem is understood.
3 marks: The maximum available mark would be awarded to a student who has demonstrated a good understanding of the chemistry involved. The student shows a good comprehension of the chemistry of the situation and has provided a logically correct answer to the question posed. This type of response might include a statement of the principles involved, a relationship or an equation, and the application of these to respond to the problem. This does not mean the answer has to be what might be termed an "excellent" answer or a "complete" one.

NATIONAL 5 CHEMISTRY 2015

Section 1

Question	Response
1.	A
2.	B
3.	D
4.	C
5.	D
6.	C
7.	C
8.	B
9.	A
10.	B
11.	B
12.	C
13.	A
14.	D
15.	C
16.	D
17.	A
18.	A
19.	D
20.	D

Section 2

1. (a) $0{\cdot}8\ cm^3\,s^{-1}$ **or** $0{\cdot}8\ cm^3/s$

   (b) Both axes labelled with units 1
   Both scales 1
   Graph drawn accurately 1
   The line must be drawn from the origin.

2. (a) Neptunium **or** Np
   or
   ^{237}Np $^{237}_{93}Np$ $_{93}Np$

   (b) Alpha **or** α **or** $^{4}_{2}\alpha$

   (c) (i) 1

   (ii) (It/Americium 241/Am-241) has a long/longer half life
   or
   will not need to be replaced as often **or** words to this effect
   or
   (It/Americium 241/Am-241) emits alpha radiation (particles) which has a low penetrating power/doesn't travel far/stopped by the smoke particles.

3. (a) (i) Hydroxyl **or** OH **or** –OH

   (ii) Ester **or** esters **or** fats **or** oils

   (b) (i) Butanoic acid
   or
   methylpropanoic acid
   or
   2-methylpropanoic acid
   or
   butyric acid

   (ii) Bromine/Br_2 decolourised/discolourised
   or
   bromine/Br_2 goes colourless

4. (a) Diagram showing two hydrogen atoms and one sulfur atom with two pairs of bonding electrons and two non-bonding pair of electrons in sulfur e.g.

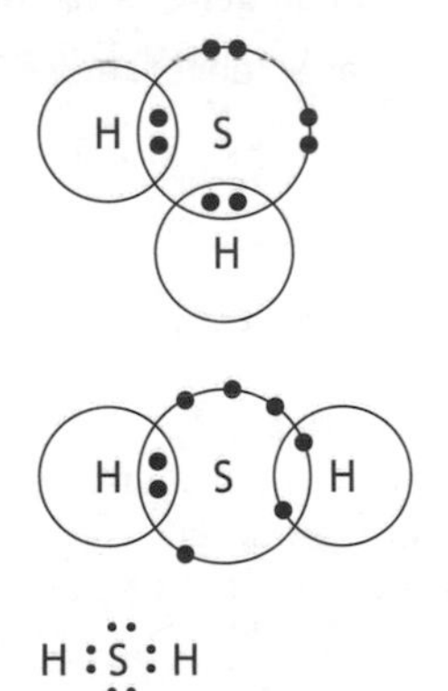

   (b) 1st = hydrogen
   2nd = hydroxide **Both required for 1 mark**

   (c) It/calcium oxide is a base
   or
   forms an alkaline solution (alkali) when dissolved in water.For the mention of alkali the candidate must explicitly state the calcium oxide is in solution/dissolved in water 1
   and
   Mention of it neutralising sulfur dioxide/it neutralises it/
   or
   a neutralisation reaction takes place. 1

5. (a) Iron **or** Fe

   (b) Any value from 52–56 inclusive

   (c) As temperature increases the yield decreases.
   or
   As temperature decreases the yield increases.
   or
   The yield increases as the temperature decreases.
   or
   The yield decreases as the temperature increases.
   Accept percentage in place of yield.

   (d) temperature 200 °C **or** a value **below** 200 °C
   and
   pressure 500 atmospheres **or** a value **greater** than 500 atmospheres **Both required for 1 mark**

6. (a) 3

   (b) $(Fe^{3+})_2(O^{2-})_3$
   or
   $Fe^{3+}{}_2\ O^{2-}{}_3$
   or
   $(Fe^{3+})_2\ O^{2-}{}_3$
   or
   $Fe^{3+}{}_2(O^{2-})_3$

or
$Fe_2^{3+} O_3^{2-}$

(c) Exothermic **or** exothermal

7. (a) Boil it
or boil off the water
or heat it
or leave it for some time/overnight/next lesson
or leave it on the window ledge
or use Bunsen (burner)
or appropriate diagram

(b) 0·2

8. (a) Method B (it)
Complete combustion/more oxygen/pure oxygen
Less/no heat loss (to surroundings)
Better insulation
Metal/platinum is a better conductor
or
Method A
Incomplete combustion
Less oxygen
(More) heat loss to surroundings
No draught shield/no insulation
Glass is a poor conductor
Flame too far away from beaker
or
Any other reasonable answer

(b) 14 /14·2/14·21/14·212

9. (a) (Metal) ore/ores

(b) (i) $4Al^{3+} + 6O^{2-} \rightarrow 4Al + 3O_2$
(or correct multiples)
All must be correct for 1 mark

(ii) Ions free to move
or
ions able to move
or
ions mobile

(c) Mg
or
magnesium
or
2Mg
or
Mg circled/highlighted/underlined in equation.

10. This is an open-ended question.
1 mark: The student has demonstrated a limited understanding of the chemistry involved. The candidate has made some statement(s) which is/are relevant to the situation, showing that at least a little of the chemistry within the problem is understood.
2 marks: The student has demonstrated a reasonable understanding of the chemistry involved. The student makes some statement(s) which is/are relevant to the situation, showing that the problem is understood.
3 marks: The maximum available mark would be awarded to a student who has demonstrated a good understanding of the chemistry involved. The student shows a good comprehension of the chemistry of the situation and has provided a logically correct answer to the question posed. This type of response might include a statement of the principles involved, a relationship or an equation, and the application of these to respond to the problem. This does not mean the answer has to be what might be termed an "excellent" answer or a "complete" one.

11. (a) 2,8,6
or
a correct target diagram

(b) $Mg(g) \rightarrow Mg^+(g) + e^-$
$Mg \rightarrow Mg^+ + e$
$Mg(g) \rightarrow Mg^+ + e^-$
$Mg \rightarrow Mg^+(g) + e$
or
$Mg(g) - e^- \rightarrow Mg^+(g)$
etc.

(c) Decreases
or
As you go from lithium to potassium (alkali metals) it (ionisation energy) decreases.
or
As you go from fluorine to bromine (halogens) it (ionisation energy) decreases.
or
As the atomic number in the group increases it decreases

12. (a) But-2-ene
or
2-butene

(b) (Molecules/compounds /hydrocarbons/alkenes) with same molecular/chemical formula but a different structural formula

(c) Correct structural formula for 3-methylpent-2-ene
or
2 ethyl but-1-ene
e.g.

```
    H       H   H
    |       |   |
H - C - C = C - C - C - H
    |   |   |   |   |
    H   H   |   H   H
        H - C - H
            |
            H
```

```
    H   H           H
    |   |           |
H - C - C - C = C - C - H
    |   |   |   |   |
    H   H   |   H   H
        H - C - H
            |
            H
```

```
    H   H           H
    |   |           |
H - C - C - C = C - C - H
    |   |   |   |   |
    H   H   |   H   H
            CH3
```

```
      H           H
      |           |
CH3 - C - C = C - C - H
      |   |   |   |
      H   |   H   H
      H - C - H
          |
          H
```

```
        H           H
        |           |
  H3C — C — C = C — C — H
        |   |   |   |
        H   |   H   H
        H — C — H
            |
            H
```

```
    H   H       H   H
    |   |       |   |
H — C — C — C — C — C — H
    |   |   ||  |   |
    H   H   ||  H   H
           CH2
```

or mirror images
or correct shortened structural formula e.g.
$CH_3CHC(CH_3)CH_2CH_3$

13. (a) Carboxyl
    (b) (i) Condensation (polymerisation)
    (ii)

```
       H   H       O   H   H   O
       |   |       ||  |   |   ||
  — O — C — C — O — C — C — C — C —
       |   |           |   |
       H   H           H   H
```

```
    H   H       O   H   H   O
    |   |       ||  |   |   ||
  — C — C — O — C — C — C — C — O —
    |   |           |   |
    H   H           H   H
```

or mirror images
Accept full or shortened structural formula or combination of both.

14. (a) (i) Carbon monoxide **or** CO/2CO
    (ii) Covalent
    (b) Distillation/distilling
    (c) The sodium **or** chlorine **or** products can be recycled/reused
    or
    Chlorine can be used in the first step
    or
    Sodium can be used in final step

15. (a) 16
    (b) 0·0032/$3 \cdot 2 \times 10^{-3}$ **or** correctly rounded answer

16. This is an open-ended question.
1 mark: The student has demonstrated a limited understanding of the chemistry involved. The candidate has made some statement(s) which is/are relevant to the situation, showing that at least a little of the chemistry within the problem is understood.
2 marks: The student has demonstrated a reasonable understanding of the chemistry involved. The student makes some statement(s) which is/are relevant to the situation, showing that the problem is understood.
3 marks: The maximum available mark would be awarded to a student who has demonstrated a good understanding of the chemistry involved. The student shows a good comprehension of the chemistry of the situation and has provided a logically correct answer to the question posed. This type of response might include a statement of the principles involved, a relationship or an equation, and the application of these to respond to the problem. This does not mean the answer has to be what might be termed an "excellent" answer or a "complete" one.

NATIONAL 5 CHEMISTRY 2016

Section 1

Question	Response
1.	D
2.	C
3.	D
4.	B
5.	B
6.	C
7.	A
8.	C
9.	B
10.	C
11.	B
12.	B
13.	A
14.	C
15.	D
16.	D
17.	A
18.	A
19.	C
20.	B

Section 2

1. (a) (i)

In the Nucleus		
Particle	*Relative Mass*	*Charge*
Proton	**1**	
Neutron		**0** **neutral** **no charge**

Both required for 1 mark

(ii)

Outside the Nucleus		
Particle	*Relative Mass*	*Charge*
ELECTRON		**-** **-1** **negative**

Both required for 1 mark

(b) 14.5
(c) (i) Pyramidal
or
trigonal pyramidal
(ii) Haber

2. (a)

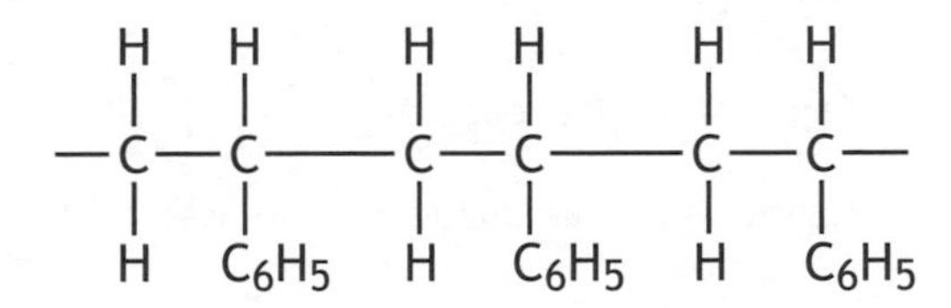

(b)

```
H   CN
|   |
C = C
|   |
H   COOCH3
```

3. (a) Exothermic
or
exothermal

(b) (i) 0·85 (with no working) **2**
Marks are awarded as follows:
$\frac{29-12}{30-10}$ or $\frac{12-29}{10-30}$
or
$\frac{17}{20}$ **(1)**
0·85 **(1)**
(ii) Any value greater than 50 and less than or equal to 60
(iii) Faster/quicker/increase/speed up

(c) (i) $Al(NO_3)_3$ circled, underlined, etc.
(ii) 0·36 (with no working) **2**
Marks are awarded as follows:
0·01 moles gives 0·015 moles **(1)**
0·015 × 24 = 0·36 **(1)**
This step on its own for 2 marks

4. (a) (i) Andalusite and kyanite
Both required for 1 mark
(ii)

Temperature (°C)	490–510
Pressure (kbar)	3·9–4·1

Both required for 1 mark

(b) 17 **or** 17·28 **or** 17·3 (with no working) **3**
With working:
GFM 162 **(1)**
28/162 × 100 (concept mark) **(1)**
This step on its own for 2 marks
Correct arithmetic **(1)**
(This mark can only be awarded if the concept mark has been awarded.)

5. (a) Unreactive
or
does not react with water/air/alkalis/(almost all) acids
or
can be beaten into shape
or
found uncombined

(b) 118

(c) (i) $CO + O_2 \longrightarrow CO_2$ **All correct for 1 mark**
or
$CO + O_2 \xrightarrow{Au} CO_2$
(ii) Catalyst **or** catalysis
or
speeds up the reaction
or
allows less energy/heat to be used for the reaction
or
lowers activation energy

(d) {hydroxide, hydrogen} {hydroxide, hydrogen} (hydrogen circled in the first pair, hydroxide circled in the second pair)
Both required for 1 mark

6. (a) (i) Flame test or correct description e.g. burn it/fertiliser/potassium, put in Bunsen flame, etc.
and
purple/lilac
Both required for 1 mark
(ii) To add/provide/supply nitrogen or it contains nitrogen. (Any wording that implies that nitrogen needs nitrogen.)

(b) $H_3PO_4 + 3NH_4OH \longrightarrow (NH_4)_3PO_4 + 3H_2O$

7. (a) This is an open-ended question.
1 mark: The student has demonstrated a limited understanding of the chemistry involved. The candidate has made some statement(s) which is/are relevant to the situation, showing that at least a little of the chemistry within the problem is understood.
2 marks: The student has demonstrated a reasonable understanding of the chemistry involved. The student makes some statement(s) which is/are relevant to the situation, showing that the problem is understood.
3 marks: The maximum available mark would be awarded to a student who has demonstrated a good understanding of the chemistry involved. The student shows a good comprehension of the chemistry of the situation and has provided a logically correct answer to the question posed. This type of response might include a statement of the principles involved, a relationship or an equation, and the application of these to respond to the problem. This does not mean the answer has to be what might be termed an "excellent" answer or a "complete" one.

8. (a) The correct structural formula for isoprene e.g.

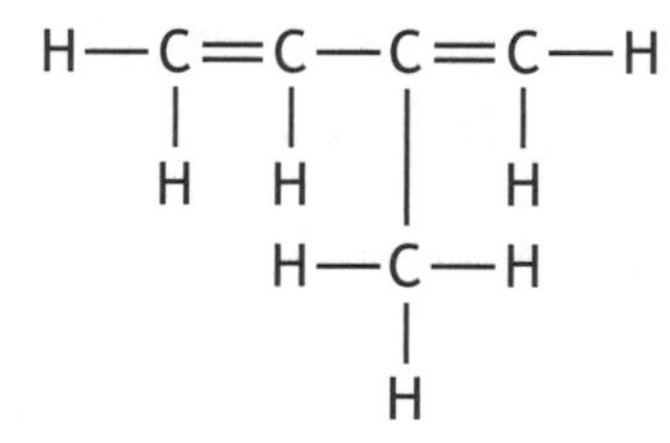

or

```
   H   CH3 H   H
   |   |   |   |
H— C = C — C = C —H
```

(However if CH_3 is used the bond must be going to the carbon.)

(b) Diagram showing delivery tube passing into a test tube which is placed in a water/ice bath.
Delivery tube must extend close enough to the neck of the test tube to ensure the vapour can enter the test tube.

(c) (i) Addition/additional
or
bromination
(ii) $C_{10}H_{16}Br_4$

9. (a) **Method A** 1

For explanation of accuracy of A (or inaccuracy of B) based on 1

- heat loss
- heat transfer
- mass loss (due to ethanol being combusted/ used up)

e.g.

- **Method A** because more heat is transferred to water
- **Method B** because less heat is transferred to water
- **B** releases more heat to the surroundings

(The second mark for the explanation will ***not*** *be awarded if the first mark is not gained.)*

(b) (i) If the alcohol is 2-ol then less energy is released compared with 1-ol or vice versa.

or

As you move from one to two (carbon/position) then the energy decreases or vice versa.

or

As it (the position of the functional group) increases/gets higher, the energy released decreases or vice versa.

or

Functional group (or it/hydroxyl/-OH) on (position) 1/end carbon/first carbon – energy released is greater/higher/bigger/increases.

or

Functional group (or it/hydroxyl/-OH) on position 2/not on the end carbon energy released is smaller/lower/decreases.

or

As it/functional group goes further along/ further down/further up the lower the energy or vice versa.

(ii) 3967–3971

(c) 55 or 55·02 (with no working) 3

With working:

Use of correct concept

$\Delta T = Eh/cm$

with **both 4·18 and 23** correctly substituted (1)

0·1 (with or without concept) (1)

Correct arithmetic (1)

(This mark can only be awarded if the concept mark has been awarded.)

10. (a) Electrolyte

(b) (i) From zinc to copper on or near the wire/ voltmeter

(ii) Ion bridge

or

salt bridge

(c) (i) Oxidation

(ii) $Br_2(\ell) + SO_3^{2-}(aq) + H_2O(\ell) \longrightarrow 2Br^-(aq) + SO_4^{2-}(aq) + 2H^+(aq)$

or

$Br_2(\ell) + SO_3^{2-}(aq) + H_2O(\ell)$

$\downarrow$

$2Br^-(aq) + SO_4^{2-}(aq) + 2H^+(aq)$

(iii) Carbon or graphite

11. (a) Methoxypropane (spelling must be correct)

(b) $C_nH_{2n+2}O$

or

$C_nH_{2n+2}O_1$

or

$C_nH_{n2+2}O$

(c)

weak
STRONG

Both required for 1 mark

(d) (i) Ethene/Eth-1-ene

or

ethylene

(ii) Any acceptable full, shortened or abbreviated structural formula e.g.

```
            H   H
             \ /
         O    C
        / \  / \
  H—   C — C    H
       /    \
      H      H

         O    CH3
        / \  /
  H—   C — C
       /    \
      H      H

           H   H
            \ /
        O    C
       / \  / \
  H2C — CH     H

        O    CH3
       / \  /
  H2C — CH
```

12. (a) Hydroxyl

(b) Weak acid, strong base/alkali

Both required for 1 mark

(c) 15·0

(d) 0·02

13. This is an open-ended question.

1 mark: The student has demonstrated a limited understanding of the chemistry involved. The candidate has made some statement(s) which is/are relevant to the situation, showing that at least a little of the chemistry within the problem is understood.

2 marks: The student has demonstrated a reasonable understanding of the chemistry involved. The student makes some statement(s) which is/are relevant to the situation, showing that the problem is understood.

3 marks: The maximum available mark would be awarded to a student who has demonstrated a good understanding of the chemistry involved. The student shows a good comprehension of the chemistry of the situation and has provided a logically correct answer to the question posed. This type of response might include a statement of the principles involved, a relationship or an equation, and the application of these to respond to the problem. This does not mean the answer has to be what might be termed an "excellent" answer or a "complete" one.

Acknowledgements

Permission has been sought from all relevant copyright holders and Hodder Gibson is grateful for the use of the following:

The extract 'Clean coal technology comes a step closer' © Russell Deeks from 'BBC Focus: Science and Technology' April 2013 (2015 Section 2 page 12);
The article 'Gold – a very useful metal' adapted from 'Education in Chemistry' Volume 45, November 2008 © Royal Society of Chemistry (2016 Section 2 page 14);
Image © Alexey Lysenko/Shutterstock.com (2016 Section 2 page 28).